Explosion Hazards in the Process Industries

Explosion Hazards in the Process Industries

Editor

Sneha Laute

Explosion Hazards in the Process Industries
Edited by **Sneha Laute**

Printed in 2017

ISBN: 978-1-68117-370-2

Library of Congress Control Number: 2015941558

© 2016 by

SCITUS Academics LLC,
616, Corporate Way, Suite 2, 4766,
Valley Cottage, NY 10989

www.scitusacademics.com

Contents

Preface

Explosions in the process industries injure or kill hundreds, if not thousands, of workers every year. They occur in process plants, refineries, platforms and pipelines all over the world. Millions of dollars are spent repairing damages, replacing equipment and rebuilding facilities in the wake of this destruction. This book explores different types of explosions that can occur in a facility and the necessary steps to guard against them. A clear set of preventative measures, rules and standards combine to make this book a convenient guide to real-world applications. Additional theoretical issues in the use of probabilistic equations and scenarios make this book an absolute necessity for process industry safety. The adjective exposible is used both in connection with dust and dust cloud, hence explosible dust and explosible dust cloud

Editor

Chapter 1

A Scheme for the Classification of Explosions in the Chemical Process Industry

Tasneem Abbasi[a], H.J. Pasman[b], and S.A. Abbasi[a]

[a]Centre for Pollution Control & Energy Technology, Pondicherry University, Kalapet, Puducherry 605014, India

[b]Mary Kay O' Connor Process Safety Center, Chemical Engineering Department, Texas A&M University, College Station, TX, USA

ABSTRACT

All process industry accidents fall under three broad categories—fire, explosion, and toxic release. Of these fire is the most common, followed by explosions. Within these broad categories occur a large number of sub-categories, each depicting a specific sub-type of a fire/explosion/toxic release. But whereas clear and self-consistent sub-classifications exist for fires and toxic releases, the situation is not as clear Vis a vis explosions. In this paper the inconsistencies and/or shortcomings associated with the classification of different types of explosions, which are seen even in otherwise highly authentic and useful reference books on process safety, are reviewed. In its context a new classification is attempted which may, hopefully, provide a frame-of-reference for the future.

INTRODUCTION

A bewildering variety of accidents occur in process industry during the storage, manufacture, and transportation of chemicals. These range from minor innocuous leaks to catastrophic releases (like the ones occurred at Seveso and Bhopal), from the barely noticed tiny sparks to all-consuming fires, and from the plop of a bubble to earth-shattering explosions of the type witnessed at Feyzin [1] and [2], Mexico City[1] and [2], Vishakhapatnam [3], Sydney [4], and Buncefield [5], among hundreds of others [1], [2], [6],[7] and [8]. The impact can also have a bewildering range, from causing temporary malfunctioning of a small component of an equipment to the demolition of an entire factory.

But, in broad terms, all process industry accidents can be classified under one or more of three categories: fire, explosion, and toxic release. Within these three broad categories fall numerous sub-categories which differ from each other in subtle as well as coarse ways. In order to forecast the likely accidents and to assess the likely consequences it is essential to properly classify the different sub-categories of accidents on the basis of their distinct attributes. Only with a proper understanding of the nature and the mechanism of each event can the consequence modelling be done effectively.

When studying the state-of-the-art of explosion modelling the authors were surprised to find that no self-consistent and/or comprehensive sub-classification of the different forms of explosions exists. Even the otherwise very authentic and exceedingly useful compendia, such as of Lees [1] and [2], CCPS[9] and [10], or the 'Yellow Book' prepared by TNO – The Netherlands Organization for Applied Scientific Research [11] – do not provide adequately clear and distinct sub-classification.

This aspect was put up for discussion by the authors during two recent international conferences, held at Pondicherry [12] and Tehran [13]. The participating scientists and safety professionals agreed that this indeed is the case.

In this paper the lacunae associated with the existing classifications of explosions are reviewed and a new classification is attempted which may, hopefully, provide a frame-of-reference for the future. This exercise assumes importance when it is recalled that of the three broad categories of accidents mentioned above, explosions cause the greatest proportion of losses in chemical process industry—an estimated 67.7% against 30.2% losses caused by fires and 2.1% by toxic releases [1] and [2].

THE EXISTING CLASSIFICATIONS

Among the most sought-after of all compendiums in the domain of process safety engineering is themagnum opus of the late Lees [1] and [2]; in it explosions have been classified as follows:

- Physical explosions
  - a. Mechanical failure of pressure system
  - b. Overpressure of pressure system
  - c. Under-pressure of pressure system
  - d. Over-temperature of pressure system
  - e. Under-temperature of pressure system
- Condensed phase explosions
  - a. High explosives
  - b. Ammonium nitrate

  c. Organic peroxides
  d. Sodium chlorate
- Vapour cloud explosions (VCEs)
- Boiling liquid expanding vapour explosions (BLEVEs)
- Confined explosions with reaction
  a. Explosion involving vapour combustion
  b. Reactor explosions
  c. Other explosions involving liquid phase reactions
- Vapour escapes into, and explosions in, buildings (VEEBs)
- Dust explosions

The problem one faces while following this classification is that its hierarchy of sets and sub-sets is unclear and overlapping. For example BLEVE, which is essentially a physical explosion, has been cited as a category separate from 'physical explosions'. Likewise vapour cloud explosions (VCEs) have been put in a different slot from 'vapour escapes into, and explosions in, buildings' (VEEBs) even though a VEEB is very similar in its mechanism to a VCE. Moreover the happenings 'c' and 'e' listed under physical explosionsare, in true sense, implosions.

Another authentic manual from a highly reputed professional body: CCPS (Centre for Chemical Process Safety) of the AIChE (American Institute of Chemical Engineers), titled Guidelines for Chemical Process Quantitative Risk Analysis [10] lists 'Explosions' as follows:

- Confined explosions
- Vapour cloud explosions
- Physical explosions
- Dust explosions
- Detonations
- Condensed phase detonations

Missiles

Curiously it lists BLEVE, which is in reality the kind of explosion which is 'louder' than most other kinds of explosions, under 'Fires'!

It is particularly surprising because one-fifth of all BLEVEs occur without causing any fire [14], [15] and [16].

In another of the oft-used CCPS manual [9], the logic diagram for explosion events given on its page 128 (Fig. 1), creates the impression that 'physical explosions' and 'confined explosions' are two distinct categories, mutually exclusive. They aren't!

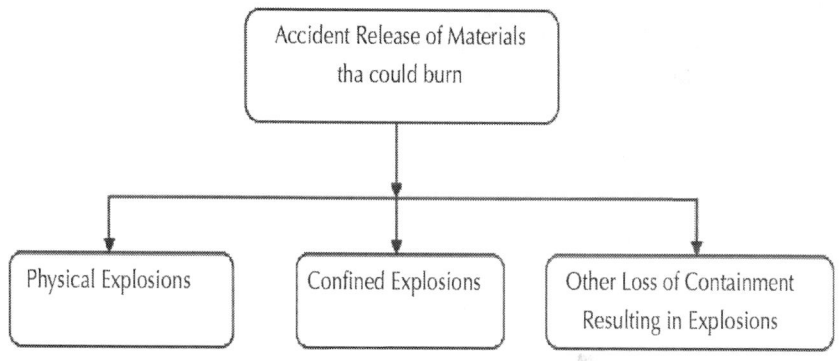

Figure 1: Logic diagram for explosion events given in CCPS [9].

Moreover in the logic diagram for physical explosion of this manual, VCE has been included among physical explosions. But VCE is not a physical explosion; it is a chemical explosion.

Numerous other texts on explosions surveyed by us, which are otherwise exceedingly useful, either provide similar forms of misclassification [17], [18], [19], [20] and [21], or no classification at all[11] and [22].

Before proceeding with a proposed new classification of explosions, it may be worthwhile to set the context by working at the definition of the term 'explosion'.

WHAT IS AN EXPLOSION?

The word 'explosion' instantaneously conjures up the image of something coming apart with a bang. In day-to-day existence we also use the word to describe any sudden burst of energy: for example a tennis player exploding into a flurry of shots; a sprinter taking off explosively; a boxer detonating his left hook. We also use expressions like as soon as the boss saw the quarterly report he simply exploded. Or the more common types: I can't tell my wife I will be late for dinner ... she will explode. The word has its origins in the Latin word explodere which means 'to drive off the stage by clapping' [23].

In more staid scientific language, the phenomenon of explosion has been defined in terms of an event's ability to generate massive overpressure. According to Lees [1] and [2], an explosion is a sudden and violent release of energy; the extent of violence depending on the rate at which energy is released. An inflated balloon or a car tyre, or a boiler heated to well above 100 °C all have energy stored under pressure. If this energy is released suddenly, it would cause a violent explosion. But if released slowly, the same extent of energy would be dissipated with no violence.

The Centre for Chemical Process Safety, American Institute of Chemical Engineers [9] defines explosion as a release of energy that causes a blast; a blast being a transient change in the gas density, pressure and velocity of the air surrounding an explosion point. Crowl and Louvar [24] make it shorter by labeling explosion as a rapid expansion of gases resulting in a rapidly moving pressure or shock wave. A highly perceptive definition has been given by F. M. Global [18]: an explosion is a rapid transformation of potential physical or chemical energy into mechanical energy and involves the violent expansion of gases.

To wit, an explosion is distinguished by the following characteristics:

- Sudden release of physically or chemically generated and stored energy.
- A shock wave/blast wave of significant magnitude, rapidly moving out from the explosion source.

Depending on the conditions of the blast, debris/flying fragments may originate from containment of the source of explosion, or materials in immediate contact with it. Cratering of the soil directly underneath the source may also lead to projectiles.

THE GENERAL PHYSICAL MECHANISM OF AN EXPLOSION

To see the relation between these various manifestations of an explosion and to better understand the differences between the various types of explosions treated in the next section, it may be helpful to recapitulate the general physical mechanism of an explosion. At the moment of explosion the energy which enables it, is present as a gas under pressure. The high pressure can be obtained by a pure, physical process (compression, heating) or by chemical conversion. The temperature of the gas may not necessarily be elevated, but often is; certainly so when the cause of high pressure is chemical conversion. 'Explosive' energy release with a rate as high as in explosions can in rare cases also take place by chemical reaction in solids, without formation of significant quantities of gas, but this kind of 'explosion' is not relevant to process safety.

The gas under pressure may instantaneously expand by bursting its containment, or may do so in the open air even when no containment is present in case its formation has been extremely rapid. Expansion of the gas causes a shock wave in the surrounding air. The mechanism of expansion can be modeled as a series of discrete compression waves which increase the pressure of the air outside the source while at the same time decreasing the pressure of the source and setting the air in motion in a direction turned away from the source (blast wind). Collision with ambient air molecules sets the latter in motion. A compression wave with its gradually rising pressure (isentropic compression) propagates with local sound velocity. This property depends on the molecule mean travel velocity and hence increases with temperature. Due to the compression in the wave the temperature of the air behind the wave increases and the subsequent waves move with higher velocities. They therefore tend to overtake the primary one, thus forming a wave characterized by a flat frontal pressure increase akin to the accelerating piston mechanism. A shock wave can be seen as discontinuous – a jump-wise – increase in pressure, temperature and material velocity, propagating through a medium with the material velocity in the direction of the front. The propagation velocity of a shock front is therefore principally higher than of a sound wave. Unlike as in a compression wave the compression in a shock wave is non-isentropic. Obviously the initial shock peak strength depends on the intensity of the source. In a three-dimensional diverging expansion of the compressed gas in free air the shock peak is followed by rarefaction in which pressure, temperature and material velocity gradually decrease and at sufficient distance from the source pressure can even become temporarily sub-atmospheric. The entire wave of propagating shock and rarefaction together is called blast (Fig. 2). At further expansion the shock peak strength decreases continually and at last, diminishes to a sound wave. If the explosion happens to be occurring near the closed end of a pipe, the resultant shock wave does not diverge and so remains planar. This prolongs the duration of the high pressure; it eventually gets weakened due to the non-isentropic nature of the compression, but very slowly. As a result the pressure decay is much slower in a pipe.

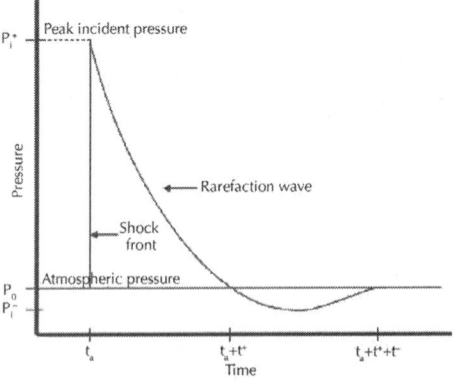

Figure 2: Blast wave led by a shock and followed by a rarefaction wave; t_a, t^+ and t^- represent times of peak incident pressure, positive phase duration and negative phase duration, respectively.

Transfer of shock wave energy from one material to the other is optimal at equal acoustic impedance (product of density and sound velocity). A shock wave in air therefore reflects most of its energy when hitting a solid or a liquid surface. An explosion in the air near the ground results by the reflection in higher pressure at the ground surface than in free air at the same distance from the source.

When the compressed gas is confined, at the moment of explosion the containment ruptures, and often is shattered. The resulting fragments are subjected to pressure difference and drag and are therefore propelled in the direction of the shock wave. There is a lot more to shock waves: their reflection and diffraction at solid (or liquid) surfaces and interactions with each other but these aspects are not pertinent to the classification of explosions.

Unfortunately it is not possible to assign precise cut-off values for the 'suddenness' of energy release, magnitude of the blast, or speed of shock wave below which an event will not be an explosion and above which it would [2], [24], [25] and [26]. This is because a lot depends on numerous other factors which may aggravate or diffuse a potentially explosive situation. For example under identical conditions of rate of pressure rise a vessel weakened at some part by corrosion or fatigue may explode while another vessel of identical rated strength may hold on. Likewise a vapour cloud meeting with some type of obstacles in its path may explode while another vapour cloud of identical material density and size meeting with different types of obstacles may not [27] and [28].

In summary, only two broad generalizations can be made:

- When the pressure build-up in any vessel or conduit exceeds the ability of the container to withstand the pressure, an explosion may result.
- When in an unconfined space, the rate of pressure rise due to the energy conversion process substantially exceeds the ability of the space to diffuse the pressure rise, an explosion may result.

TYPES OF EXPLOSIONS

As mentioned earlier, one encounters a wide variety of explosions, in terms of nature as well as magnitude, ranging from what we get if a water droplet accidentally falls on hot oil in the kitchen, to firecrackers; a bursting bubble to earth-shaking blasts; a 0.22 gunshot to nuclear bombs. But, in essence, only three kinds of energy are associated with all explosions: physical, chemical, and nuclear. Of these, only the first two are encountered in process industries as also in day-to-day existence. We would, therefore, dwell further only on these two. In addition this classification being specific to explosions which occur in chemical process industry in the course of process operations, and during the storage and transport of associated chemicals, we have omitted external happenings such as a lightening strike, sabotage, earthquake, etc., which may cause an explosion. Hence the first order of the proposed classification (Fig. 3) comprises of two categories: physical explosion and chemical explosion.

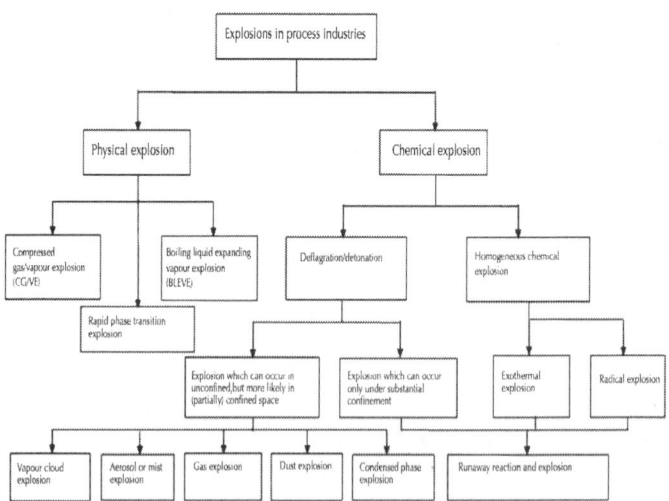

Figure 3: The proposed scheme for the classification of explosions in chemical process industry.

PHYSICAL EXPLOSION

Before proposing a definition of 'physical explosion' we must emphasize that in all types of explosion much physics is involved. In fact the gas expansion process explained in the previous section is purely physical, but even in an explosion in which the energy is generated in a chemical conversion the mechanism of propagating the reaction is by heat transfer or compression and hence physical.

The term 'physical' in physical explosion refers to the way the energy that enabled the explosion would have been accumulated. It could have been by heating a volatile liquid or a gas in a containment and thus causing pressure build-up. Or mechanically, by simply compressing a gas. The accumulation process may be relatively slower than in the 'chemical explosions' defined in Section 6. A physical explosion occurs when the accumulated energy is suddenly released in a rapid physical change such as the expansion of a compressed gas or the flash vapourization of a superheated liquid by a failure somehow of the containment. After the explosive release a substance may undergo chemical reactions, a flammable substance may start burning due to mixing with air contributing heavily to the overall effect as with the failure of a tank of compressed, liquefied hydrocarbons, but the cause of all physical explosions is purely mechanical energy. It may be mentioned that electrical discharge can also cause a shock/blast wave in surrounding air, as with a lightning strike, but this form of physical explosion is not considered here.

Depending on the situations that cause physical explosion such an event may be categorized as follows.

Compressed Gas/Vapour Explosion (CG/VE)

CG/VE refers to catastrophic rupture of a pressurized gas-filled vessel. The vessel may be filled up entirely with the gas or may contain some liquid. In the latter case we call it CG/VE only if the liquid is not in a superheated state at the instant of vessel failure;

in other words when it is at a temperature below its atmospheric pressure boiling point.

A burst of a pressurized vessel in CG/VE can occur for the following reasons:

- Failure of level control, or of pressure regulating and pressure relief equipment (physical over pressurization).
- Reduction in vessel wall thickness due to:
  a.   corrosion,
  b.   erosion,
  c.   chemical attack.
- Reduction in vessel wall strength due to:
  a.   overheating,
  b.   material defects with subsequent development of fracture,
  c.   chemical attack, e.g., stress corrosion cracking, pitting, embrittlement,
  d.   fatigue induced weakening of the vessel.
- Any other mechanical cause (such as a container suffering injury due to accidental fall or a hit).

In situations as in 1, above, failure can normally occur only when the vessel acquires pressure significantly higher than the operating pressure. In other situations failure can occur even at or near the operating pressure of the vessel. In a situation as in 3a, above, accidental overheating of the vessel may cause some or all of its liquid contents to vapourize, thereby taking the internal pressure beyond the tolerance level of the vessel. This may lead to a variant of CG/VE called rapid phase transition explosion.

Boiling Liquid Expanding Vapour Explosion (BLEVE)

A BLEVE differs from a CG/VE explained above on two counts:

- A vessel suffering a BLEVE must contain significant quantities of the substance in a liquid form.

- The liquid should be existing at a temperature which is well above the liquid's boiling point at normal (atmospheric) pressure, in other words in a 'superheated' state relative to the normal pressure.

If a vessel containing a superheated liquid under pressure is suddenly depressurized, the liquid suffers instantaneous nucleation and flash vapourization into gas several times larger in volume than the parent liquid [27] and [29]. The resultant build up of pressure may force the vessel to fail catastrophically, causing a BLEVE. Hence, whereas in CG/VE (described in the preceding section) the bust of energy to generate the explosive pressure wave comes solely from the adiabatic expansion of a gas, in BLEVE the adiabatic vapourization of liquid is the main contributory factor.

For risk analysis distinction is made between a 'hot' and a 'cold' BLEVE, depending on whether the vessel fractures/punctures by mechanical loading or by an external fire. In the latter case the effects can be more severe due to heat input and higher pressure at fracture.

A BLEVE gives rise to the following [4], [14] and [29]:

- Blast wave.
- Flying fragments (missiles).
- Splashing of some of the liquid to form short-lived pools; the pools may be on fire if the liquid is flammable.
- Fire or toxic gas release. If the pressure-liquefied vapour is flammable, as is often the case, the BLEVE leads to a (rising) fireball or at delayed ignition (rarely) to flash fire or vapour cloud explosion. When the material undergoing BLEVE is toxic, as in the case of ammonia or chlorine, adverse impacts include toxic gas dispersion.

Rapid Phase Transition Explosion (RPT)

Rapid phase transition (RPT) explosion may occur when cryogenic liquids are accidentally exposed to hotter environment, for example liquefied natural gas (LNG) spilled on or in water [29] and [30]. The precise conditions for the local instability to cause a RPT are still unclear; threshold amounts of ethane and propane in the natural gas are a factor. The effect is relatively weak. The largest blast measured is that of 3.5 kg TNT equivalent. The main effect is producing a large puff of evaporated natural gas temporarily increasing in cloud size.

A variant of RPT which is rare in chemical process industries but more common in metallurgical industries occurs when molten metals or hot oil accidentally come in contact with a much cooler water (or other liquid). This variant, RPT and BLEVE have one aspect in common—in each heat energy is suddenly and abundantly transferred to a liquid which superheats and thereby undergoes instantaneous nucleation. There is a burst of 'boiling liquid and expanding vapour'. The most common, and by far the most destructive, manifestation of this aspect occurs in the form of BLEVEs in vessels containing pressure liquefied gases.

CHEMICAL EXPLOSION

When the slug of energy needed to generate large quantities of gas within a very short time span, leading to a rate of pressure build-up that is fast enough to cause an explosion even in open space, comes due to a chemical reaction, we may call such an event chemical explosion.

Homogeneous Chemical Explosion, Deflagration and Detonation

The broadest sub-classification of chemical explosion proposed by us is based on where in the material the reactions are taking place at a given time. It can happen in two distinct forms: if it is occurring throughout the mass of material all at once, the phenomenon may be called a homogeneous chemical explosion. In case the reaction occurs only in a propagating reaction zone it can be in two well-defined but very different intensities: deflagration and detonation. The latter can be made visible as moving flames; the velocity in case of deflagration ranges from very low up to some hundreds of meter per second, while in detonation it is of the order of kilometers per second.

The broad sub-classfication is explained further in Section 6.5. Under the second of this broad category – deflagration/detonation – fall chemical explosions which can either occur only under substantial confinement or the ones that can also occur in unconfined space.

Homogeneous Chemical Explosion

A homogeneous chemical explosion can occur in two ways: sharp rise in temperature due to an exothermal chemical reaction or due to the formation of a net surplus of radicals.

We propose to call them exothermal explosion and radical explosion respectively, although the former usually is loosely called thermal explosion [1], [2] and [31].

The name exothermal explosion hints at the acceleration mechanism in the reactions. To produce hot gas fast, reactions are needed which overall are exothermic. The heat of reaction is partially lost to the surroundings but partially increases the temperature of the reactants and consequently accelerates the reaction. If no heat is lost (adiabatic situation) or the medium is perfectly stirred, the temperature is equal throughout and ideally the reaction rate is also equal throughout. We propose to call this extreme ahomogeneous exothermal explosion. Another mechanism may also be operative: made possible by branched radical reactions. Reactions propagate often by radicals; hydrogen–oxygen or hydrogen–chlorine reactions are well-known examples. If a reaction occurs in which more radicals are produced than necessary for the reactants to be formed from the original substance, there is 'branching'. This can accelerate the overall reaction even at relatively low temperatures. Peroxide intermediates formed in the reaction of hydrocarbons with oxygen decompose that way. In gases these compounds cause so-called 'cool flames', which are not always harmless. It must be said that the two forms – exothermal and radical explosion – cannot always be sharply distinguished.

Deflagration and Detonation

Deflagration literally means 'fast burning'; it consists of a moving exothermic reaction zone sustained by heat flow from hot reaction gases to unreacted material by conduction, convection, and radiation [1], [2],[29] and [31]. Propagation can become very slow when unconfined (for example in some ammonium nitrate fertilizer formulations it can be just a few centimeters per hour) to extremely fast; up till hundreds of meters per second. But it is always subsonic with respect to the sound velocity in the material.

In a detonation the energy transfer to initiate a reaction in a fresh substance is caused by compression in a shock wave (reactive shock), hence the propagation velocity is supersonic [1], [2], [23] and [27]. Because of the high velocity of the wave and hence the very short time within which the reactions and the conversion of the substance to a hot expanding gas mass takes place, a shock wave is produced in the ambient air which we perceive as a 'bang'.

The temperature increase by the sudden compression starts the reaction. In condensed substances detonation velocities can, in extreme cases, reach 10 km/s, creating hundreds of megabars in pressure. In gases, on the other hand, the sound velocity is much lower so is their detonation velocity, up to 3 km/s. Due to the lower density of gases the pressures are also lower by factors of ~1000. Whether on the higher side of intensity or lower, due to the high peak pressure detonations are always very destructive.

The high propagation rates in deflagrations occur either under confinement by a mechanism of pressure driven acceleration due to increased heat transfer and higher reaction rate, or in gases by flame acceleration as a result of turbulence generation in the still unreacted gas in front of the flame. The turbulence increases the burning surface area and 'thickens' the flame. It also, to a certain extent, increases the burning rate relative to the substance. In grained solids with some confinement the hot reaction gases may penetrate into the mass, spreading ignition, and this may very quickly increase the pressure to high levels.

The mechanisms mentioned above may lead to transition between deflagration and detonation (DDT) [2],[32] and [33] which can occur in gases, liquids and solids, leading to catastrophic pressure effects. In the transition stage, the pressure may be temporarily even higher than in a stationary detonation wave. The mechanism of the transition is rather complex but can be explained in rough terms as follows. If due to the acceleration, the deflagration flames reach velocities of hundreds of meters per second, the accompanying compression waves may grow very strong. Either before or just behind the flame front in the local compressed substance, conditions may become such that any perturbation in the reacting mass may be strengthened to a shock wave. It may then suddenly propagate as a detonation wave ('explosion within an explosion').

Deflagrations are often accompanied by flame, detonation always. In both cases reaction products often burn after being mixed with air but this does not contribute to explosion dynamics.

Fuel and Oxidiser Mixtures

As explained above many chemical explosions occur in mixtures of an oxidiser and a fuel. This is true for solids, e.g. pyrotechnics; and liquids, e.g. certain types of liquid explosives; but particularly true for mixtures of a fuel and air oxygen as oxidiser. These form the basis for a distinction of explosions used much in practice based on the phase or aggregation state of the fuel:

- Gas and vapour fuelled—gas explosion/vapour cloud explosion.
- (Combustible) dust fuelled—dust explosion.
- Aerosol (in the form of liquid droplets) fuelled—aerosol or mist explosion.

Quite often what is reported as a vapour cloud explosion is either a gas explosion in the open or an aerosol explosion. Most hydrocarbons when released accidentally from a pressure-liquefied or refrigerated condition are initially cold and heavy. They either partly condense by themselves or cause the condensation of the water vapour in the air in which they mix. They can be seen just before the explosion as a white cloud, lying low. A notorious example of an aerosol/vapour cloud explosion is the one that occurred in December 2005 at the Buncefield fuel depot in the U.K. A spill of gasoline from an overflowing tank formed a flammable aerosol which spread over a wide area before it ignited, exploded and set fire to a considerable number of other tanks [5].

In deflagrations and detonations of mixtures of fuel and oxidiser energy release rates and therefore violence of explosion vary with composition. Certain compositions can be found where explosion just does not occur anymore. These compositions form the explosion limits and in between is the explosion range. In case of gas or dust explosion the US terminology uses 'flammability limit' and 'flammable range'; the 'flammability limit' is expressed in terms of mean composition of the fuel–oxidant mixtures in which flame propagation is just possible. In Europe the expressions 'explosion limit' and 'explosion range' are in vogue and the composition which just fails to ignite is declared as limit. Detonation ranges are narrower than deflagration ones. Deflagration near the explosion limit ensues only an upward propagating flame.

Degree of Confinement

Further classification of chemical explosion is proposed as follows:

- Explosion which occurs in unconfined, partially confined, as well as confined space.
- Explosion which can occur only with the substance reacting under confinement (since otherwise the reactions are of too low intensity or too slow).

The first category is specific to chemical explosion because no physical explosion in chemical process industry (in other words excluding nuclear explosion) is known to occur in unconfined space. Under this category fall the following types of explosions relevant to process industry; these quite often involve mixtures of an oxidiser and a fuel:

- Condensed phase explosion (such as the ammonium nitrate explosion in Toulouse, France in 2001;Table 1).
- Vapour cloud explosion
- Aerosol/mist explosion
- Gas explosion
- Dust explosion

A condensed phase explosion does not need any confinement. But for the vapour cloud or the aerosol explosions to occur, some degree of congestion or confinement is necessary so also the presence of significantly high quantities of flammable substance. These twin conditions provide sufficient fuel and feedback turbulence to the flame for attaining high enough flame speed leading to a blast. Gas and dust explosions usually take place inside partially confined or vented equipment.

How the physical state of a substance influences the nature of the explosion it undergoes is typified by condensed phase explosions, too. It makes a difference whether a substance is in the form of fine grains or a solid block, whether it is a liquid, a mix of a liquid and a solid, or a liquid and a gas. Usually mixed aggregate states allow physical interactions of shock and flame leading to a more violent reaction than single phase substances. Aggregates may also attain transition from deflagration to detonation more easily. Episodes have occurred in the past when, upon being heated by accidental fires, energetic materials have undergone detonation even in open space.

In the second category come runaway reactions mostly of liquids (reactor explosion) and solids (self-igniting exothermic decomposing substances or combustibles reacting with air at elevated temperature). These may occur with or without rapid phase transition causing exothermal explosion, sometimes partially resulting in deflagration. In rare cases the deflagration may transit into detonation. An exothermal explosion in a fuel–oxidiser gas mixture can also occur, at initially elevated pressure. After the explosion of a reactor, the ejected flammable products or reactants can mix with air in the space above the reactor forming a hot cloud which may self-ignite causing a vapour cloud explosion above the reactor.

Explosion Intensity and Ease of Initiation

As mentioned, the intensity of the phenomena which leads to chemical explosion, may vary considerably. It distinguishes what we may call 'milder' chemical explosions from the more severe ones. It applies in some way to all categories of chemical explosions. Substances that react rapidly and violently (high energy of activation and heat of explosion) produce relatively intense pressures. They are able to propagate the explosive reaction even in strong diverging geometry and hence can be initiated relatively easily by point source. Usually the required overall energy can be relatively low. Less reactive substances need energetic initiation by either a point source of larger diameter or by initiation over a larger plane area of the substance. In the later case the source itself can also be of lower intensity.

The energy producing reactions in all chemical explosions can be in a pure substance between atoms from the same molecule (intramolecular), or in a mixture of different substance components (intermolecular). The energy release rate of the latter type is usually lower and hence can determine whether a substance can only deflagrate or also detonate. Slower reaction rate also explains why sometimes absence of confinement can abort an explosion.

In summary three dimensions or facets influence the nature and severity of chemical explosions: the physics of propagation (exothermal explosion, deflagration, detonation), the aggregation state of the fuel–oxidiser mixture (gas, dust, aerosol, or condensed phase) and the presence or absence of confinement. If proper attention is not paid to the existence of all these dimensions and the way in which they influence each other, confusion and misinterpretation occurs leading to flaws in the accident modelling as well as in subsequent control strategies.

Distinctive Features of Specific Explosion Types

The distinctive features of several forms of deflagration/detonation falling under the two categories mentioned in Section 6.3 are presented below under their most common names. Attention is also paid to the processes of explosion initiation.

Vapour Cloud Explosion (VCE)

When large quantity of flammable vapour or gas is accidentally released into air it may form a vapour cloud. If the release is from a pressure liquefied state, its initial behaviour may be similar to that of a heavy gas even if at normal temperature and pressure the substance may be lighter-than-air. This may be due to its initially low temperature, entrapped liquid droplets (condensed fuel vapour or, in case of high humidity, water mist), and high release density. The resultant vapour cloud is, therefore, likely to hug the ground, at least initially, before slowly rising and moving. In case no immediate ignition resulting in jet flame (torch), (stratified layer) flash fire or rising fire ball occurs the vapour cloud disperses. At sufficiently delayed and strong ignition a vapour cloud explosion (VCE) may take place. Because of strong blast VCEs have the potential to cause heavy damage [17], [20], [21] and [22].

Once flammable material has been accidentally released, for it to lead to a VCE, the following requirements must be met with [2] and [28]:

- The released material must be flammable.
- The vapour cloud must mix with air to produce, depending on the fuel's reactivity, a sufficient mass in the flammable/explosive range of the material released.

- The environment of the cloud should offer sufficient confinement/congestion for turbulence generation in the flow driven by expanding hot combustion gases to initiate a feed-back flame acceleration process and pressure wave reflections.

Initially the flame ball expands from the place of ignition but the front propagates slowly with respect to the unburned gas, unable to produce a blast. However, while growing, the flame front stretches, starts wrinkling at the surface due to instabilities, and energy release starts to accelerate. Flame surface further increases; in particular passing around obstacles to the flow; further distorting the surface but also generating turbulence in fresh gas pushed ahead; resulting in turbulent flame brush and higher velocities. This feed-back mechanism characterizes VCE and produces its blast.

Analysis of past accidents has revealed [17], [27], [28] and [29] that partial confinement as may occur when the vapour cloud develops in or over, or drifts towards structures such as plant machinery, pipe racks, tanks, buildings, or vegetation all strongly accelerate the flame. Such a rapidly accelerating flame with the hot expanding gas behind it acts like a porous but accelerating piston and generates compression, shock and blast. The high damage potential of such a blast is due to the overpressure and impulse effect of the blast wind on objects. If there were no flame acceleration, the damage would have confined to thermal radiation and direct flame impingement.

A VCE to occur and result in a blast in a totally unconfined space is rare but it can happen with a very strong initiation source such as a detonating high explosive charge.

Following sub-classification of VCE can be done:

- VCE occurring in unconfined space.
- VCE occurring in hollow (semi-)confinements (such as empty spaces in a building).
- VCE occurring in a relatively open space where drag-generating obstructions to the flow of gases ahead of the expanding flame ball are present (congested area).

Aerosol (or Mist) Explosion

An aerosol explosion is quite similar to a vapour cloud explosion; the difference occurs in the role played by the liquid droplets contained in an aerosol. Their presence enhances the probability of the cloud getting into the flammable range. Secondly once a flame is initiated it generates a blast which in turn can interact with particles ahead of the flame stripping off a micro-mist due to drag. This fine mist can react very intensively when it is reached by the flame. This can make the explosion more violent.

Gas Explosion (GE)

'Gas explosion' is a name classically given [34] to explosion which takes place in a container or a conduit that happens to carry a fuel and an oxidiser (both in gas phase) in a mixture ratio within the explosion limits.[1] If the initiation is by a strong shock as from a very energetic spark, a high explosive charge, an exploding wire or a focussed laser locally ionising the air; and if the mixture is sufficiently reactive; an immediate detonation my occur. Otherwise, at least initially, there is deflagration. When the containment is able to sustain the explosion pressure not much is observed outside the containment; at best a sound ensues. Due to pressure oscillations generated by a rising flame instability, this sound can become rather strong.

Additional hazards occur when a gas explosion takes place in a space consisting of chambers inter connected by pipes. An explosion in one of the spaces causes increase in pressure of unburned gas in others. When the flame reaches these spaces the resulting explosion starts at a higher pressure level. This effect keeps propagating, leading to 'pressure piling'. The final pressure is substantially higher than of the initial explosion, with correspondingly high damage potential. Another effect is the flame acceleration that can occur in pipes due to turbulence generated by friction of the flow ahead of the propagating flame. Such acceleration can lead to very violent deflagration propagating with hundreds of meters per second. Over a sufficient run-up distance it can metamorph into a detonation generating much higher pressures. Further enhancement of turbulence, hence flame acceleration, can be caused by pressure waves which may occur due to flame instabilities or sudden opening of avent. The later may cause the waves to reflect against solid surfaces and repeatedly pass through the flame zone, exacerbating the effect.

A great deal of what has been described above holds true for combustible aerosol–oxidiser mixtures (such as dusts in air and fuel sprays) as well. One fundamental difference is that the behaviour of those mixtures cannot be studied in a stagnant flow field; turbulence to some extent is essential to maintain the aerosol condition. Because of its distinct characteristics, dust explosion is treated below as a category of its own.

Dust Explosion (DE)

When combustible, dust-sized particles of a flammable material get (accidentally) suspended in air and the resulting dust cloud catches fire, a dust explosion may result. Ignition may be by a variety of sources such as open flame, mechanical and electrical sparks, friction or other type of heating, such as by an unprotected lamp or even by self-heating of the dust settled in a layer in e.g. a dryer. The reactivity of a dust increases, up to a limit, with the decrease in particle size, increase in surface area to mass ratio, decrease in moisture content, and increase in combustion energy. Explosion indicators are of the same order of magnitude as those of explosive gases, except that the upper explosion limit may be much higher. As in case of a gas explosion, dust explosions generally occur within confined or partially confined space but can do so, albeit rarely, in relatively open spaces as well [6] and [29].

Therefore dust explosions (DE) can also be sub-classified as:

- DE occurring in a confined and vented space such as machinery for diminution, mixing, drying, granulating, separating, filtering, transportation piping, Jacob ladders or storage (silo).

- DE occurring in hollow semi-confinements (such as empty spaces in a building, or in corridors as in coal mines and conveyor belts).

- DE occurring in unconfined space when the space contains sufficient dust both settled and dispersed in the air and such obstructions to the flow that the velocity of the flame remains sufficiently high to feed itself by whirling up further dust.

An additional hazard is associated with the layers of combustible dust lying around on floors, equipment, etc. It may whirl up when a weak local blast occurs by a starting dust explosion. The explosion then progresses by the whirled up dust and so feeds itself. Also dust explosion flames can accelerate in sufficient confinement and over sufficient run-up distance and such acceleration can eventually lead to detonation. Metal dust explosions are renowned by their violence and high temperatures causing intense heat radiation. Depending on the type of metal and oxygen content in the mixture, self-ignition may also occur. Protection from such eventuality may be ensured by compartmentalization which may restrict the flame from passing on other parts, and by extinguishing, venting, etc. [6] and [35].

Condensed Phase Explosion

Certain industrial liquid or solid products of high energy density – also called energetic materials' – on catching fire, can generate pressure waves of energy and speed high enough to cause an explosion even in an unconfined space. But, as with VCE and dust explosion, usually confinement enhances the ferocity and damage potential of condensed phase explosions. Ammonium nitrate, sodium chlorate, organic peroxides are among the industrial chemicals associated with condensed phase explosions. Their reactions are similar but usually less energetic and less complete than of high explosives, propellants and pyrotechnics. Because of the lower reactivity of many of these industrial chemicals an initiation by a point source or a fire at first results in deflagration. Given sufficient mass and confinement this can then turn into a detonation. Due to the presence of impurities/contaminants or moisture some of the condensed phase explosions may be restricted to a deflagration.

In case of substances with low energy density, point initiation must be strong enough to initiate reactions to such an extent that it results in sustained, stable detonation. Initiation over a larger surface area, for example with an explosive pellet, generally results in a stable propagating reaction. Part of the energy in the slow reacting substances, in comparison to high explosives, is released too far behind the shock front to enable sustaining of the detonation wave front. Small grain size, edgy shape, catalysing traces, large porosity, etc., enhance detonability.

Some low-reactive substances only explode when under confinement in a steel vessel or otherwise. There is a run-away reaction following a pre-heating process; it then builds up to a deflagration.

A special but rare form of explosion is low-velocity-detonation in which a shock wave generated by a reaction in a substance propagates ahead of the reaction zone in a contact material possessing a higher sound velocity such as steel. Pressure waves radiating back in the substance pre-heat and so a higher propagation rate of the zone occurs than would have without the contact material. Some nitro-alkanes and -aromats are examples.

Runaway Reaction (Reactor Explosion) With or Without Phase Transition

A runaway reaction is a chemical reaction in a gas, liquid or solid material which accelerates out of control as a result of the heat of reaction or decomposition exceeding the cooling capacity of the containing vessel[2] and [36] or by the multiplying effect of radical reactions. Hence initially it is a form of exothermal or of radical explosion. It may be clarified that radical decomposition mechanisms may often play a role but it need not be the production of net more radicals as in cool flames or in decomposing organic peroxides. Had runaway reactions been occurring throughout the substance in a homogeneous fashion, and everywhere at the same rate, their description would have been relatively simple. In practice, due to physical causes (heat loss, convection) and chemical ones (local concentration gradients, catalysing products) in most cases the situation is far from homogeneous. The pressure increases due to production of heat, formation of non-condensable gases, and/ or an increasing vapour pressure of liquid components. In a final stage (cook-off) the substance may deflagrate with a reaction zone starting at a hot spot and propagating through an already reacting substance. In rare cases the process may develop into a detonation. Near the end of the explosive run-away process, the situation is thus often complex and unclear as regards the intensity of the explosion.

This process can only be controlled by a properly dimensioned pressure relief device, like a rupture disc or pressure relief valve. If these devices malfunction or if the reaction is faster than foreseen, the pressure may continue to rise until the vessel fails [1], [2] and [36].

In some cases the contents of the vessel may decompose within milliseconds and the reaction of the entire contents may even be completed before the vessel bursts open [37]. The explosions they cause are more like explosions from high explosives than like pressure vessel bursts.

An exothermic runaway reaction can render the reactants superheated and if the pressure build-up exceeds the tolerance level of the process vessel, the resultant explosion may be similar to a BLEVE [4]. But we would distinguish such explosions from BLEVE on the following counts:

- The energy involved in the build up of temperature and pressure came from chemical reactions.
- The contents were not in the pressure-liquefied form from the outset.
- The explosion was not caused by the accidental weakening of the pressure vessel but rather from the excessive pressure generated due to exothermic runaway reactions.

Vapour cloud explosion	1. Flixborough, UK, 1974: One of the most extensively documented of all vapour cloud explosions occurred when cyclohexane was accidentally released from a ruptured bypass in a reactor train and found a source of ignition. The resulting blast and the fires destroyed not only the cyclohexane plant but several other plants, too.	[1] and [2]
	2. Ufa, USSR, 1989: A leak in the Trans-Siberian LPG pipeline, which had gone undetected for several hours, formed a vapour cloud which extended five miles in one direction. Two trains, travelling in opposite direction in railway lines passing nearby, provided the spark that led to a massive vapour cloud explosion which killed 462 persons.	[2] and [42]
Aerosol or mist explosion	1. Phillipsburgh, NJ, USA, 1959: Oil mist in a compressor test facility exploded, killing 6 and injuring 30.	[29]
	2. Sirdal, Norway, 1973: Oil mist explosion in a transformer room killed 3 and injured several others.	[29]
Gas explosion	1. Newham, East London, 1968: A gas explosion on the 18th floor of a 23 storey high-rise pushed out the load bearing side wall of a floor causing a progressive collapse of the entire corner of the block.	[1] and [2]
	2. Netherlands, 2003: An explosion in a gas-fired furnace at a melamine plant killed 3. The explosion occurred at a start-up after a maintenance operation.	[43]
Dust explosion	1. Harbin, People's Republic of China, 1987: Electrostatic spark in one of the seven linen dust collecting units led to a massive explosion which destroyed all seven units and most of the rest of the plant; killing 58 and injuring 177.	[6]
	2. Kinston, NC, USA: a cloud of polyethylene dust in a pharmaceutical factory exploded, killing 6 and injuring 38.	[6]

Condensed phase explosion	1. Toulouse, France, 2001: A storage facility that held 200–300 tonnes of ammonium nitrate suffered a massive condensed phase explosion that created a 7 m deep crater, killed 29 and injured 2500.	[2]
	2. Ryongchon, North Korea, 2004: Possibly an electric short-circuit triggered a condensed phase explosion in a goods train loaded with ammonium nitrate. The impact of resultant blast covered a radius of ~2 km, generating a 15 m deep crater and destroying 8100 horses. Over 150 persons died and 1300 were injured.	[43]
Radical explosion	1. Gifu, Japan, 1971: A vinyl acetate monomer, which was erroneously stored in a 1 m³ drum without adding a polymerization inhibitor, went through spontaneous polymerization causing build up of temperature and pressure. Eventually there was an explosion and concomitant fire which caused extensive damage.	[44]
	2. During start-up of a chlorination plant some problems arose and the start-up was terminated after chlorine and an organic material had been fed to the reactor. As no clear-cut purging procedure for this situation had been advised, after handling the problem the start-up was repeated. A short while later the reactor exploded.	[45]
Runaway reaction	1. Stanlow, Cheshire, 1990: A 15 m³ batch reactor at a Shell plant set to produce 2,4-difluoro-aniline had a runaway reaction leading to an explosion so severe that the vessel body unwrapped into a flat plate. The cover was hurled up 200 m and other missiles reached up to 500 m. The entire plant was devastated and nearby buildings also suffered structural damage.	[2]
	2. Institute, WV, USA, 2008: Runaway reaction created extremely high temperature and pressure in a vessel at a Bayer factory, causing an explosion which demolished process equipment and ruptured pipes and conduits. Two operators were killed and eight suffered from toxic inhalation.	[46]

Rapid phase transition	The phenomenon has been observed in spill experiments with LNG on water such as in the Coyote test series conducted at Lawrence Livermore National Laboratory and Naval Weapons Center at China Lake, USA in 1981.	[47]

Some of the well-known terms like 'dust explosion' and 'gas explosion' featuring in the proposal classification are not quite precise. Yet they have been retained because the terms have been in vogue since long and are now firmly entrenched in the process safety lexicon. Moreover, unlike some other terms which are used by different authors in different sense to denote a same type of explosion, these terms connotate phenomena of which interpretation is fairly consistent across different users.

Perhaps, in due course, terms like 'dust explosion' – which, to the uninitiated, would imply that it involves exploding dust rather than what it really is, viz a combustible dust reacting explosively with the surrounding oxidant – be substituted with more correct and precise terms. Likewise a better name for gas explosion – which otherwise may be similarly misleading to a lay person – may be found. But it has to be an initiative to be taken globally by a consortium of a large number of representative bodies. Till such time it happens the systematization proposed here can be used as a starting point to set that process of a consensus-based systematization in motion. The relative degree of specificity, hopefully achieved by the present attempt, is also bolstered by the fact that the models which predict the causes and effects of explosions as classified now are clearly distinct from each other. Moreover the measures needed to prevent these different types of accidents, or to cushion their impacts if the accidents do occur, are distinct from each other due partly to the different aggregation states and partly to the different degrees of confinement.

There may be situations wherein corrosion may gradually weaken some part of a vessel containing a pressure liquefied gas till, at one point, the vessel suffers a BLEVE. A question may arise whether to call it a chemical explosion or a physical explosion?

Likewise, if a vessel weakened at some part on account of creep or fatigue explodes due to a runaway reaction, should it be classified as a physical explosion?

The classification presented by us is based on the source of energy that is involved in an explosion and is independent of the factors which may make a process unit vulnerable to an accident. By this criterion the first of the abovementioned instances will be a physical explosion and the second one a chemical explosion.

ACKNOWLEDGEMENTS

Authors thank the Chemical Engineering Programme, Department of Science & Technology, Government of India, New Delhi, for support.

REFERENCES

1.   F.P. Lees, Loss Prevention in the Process Industries—Hazard Identification, Assessment, and Control, vols. 1–3, Butterworth-Heinemann, Oxford, 1996.

2.   F.P. Lees, in: S. Mannan (Ed.), Loss Prevention in the Process Industries—Hazard Identification, Assessment, and Control, vols. 1–3, Elsevier/ButterworthHeinemann, Oxford, 2005.

3.   F.I. Khan, S.A. Abbasi, The world's worst chemical industry accident of 1990s—what happened and what might have been. A quantitative study, Process Safety Progress 8 (1999) 135–141.

4.   T. Abbasi, S.A. Abbasi, The boiling liquid expanding vapour explosion (BLEVE): mechanism, consequence assessment, management, Journal of Hazardous Materials 141 (2007) 489–519.

5.   http://www.buncefieldinvestigation.gov.uk, Last accessed on July 29, 2008.

6.   T. Abbasi, S.A. Abbasi, Dust explosions—cases, causes, consequences and control, Journal of Hazardous Materials 140 (2007) 7–44.

7.   F.I. Khan, S.A. Abbasi, DOMIFFECT (DOMIno eFFECT): user-friendly software for domino effect analysis, Environmental Modelling and Software 13 (1998) 163–177.

8.   F.I. Khan, S.A. Abbasi, An assessment of the likehood of occurrence, and the damage potential of domino effect (chain of accidents) in a typical cluster of industries, Journal of Loss Prevention in the Process Industries 14 (2001) 283–306.

9.   CCPS: Centre for Chemical Process Safety, Guidelines for Consequence Analysis of Chemical Releases, The American Institute of Chemical Engineering, New York, 1999.

10. CCPS: Centre for Chemical Process Safety, Guidelines for Chemical Process Quantitative Risk Analysis, The American Institute of Chemical Engineering, New York, 2000.

11. C.J.H. van den Bosch, R.A.P.M. Weterings, Methods for the Calculation of Physical Effects: 'Yellow Book' CPR 14E (pt 2), Committee for the Prevention of Disasters, The Hague, 1997.

12. S. Sundaramoorthy (Ed.), Proceedings of the International Conference on Cleaner Technolgoies, Allied Publishers, Chennai, 2007.

13. D. Rashtchian (Ed.), Proceedings of the Second National Conference on Health, Safety, and Environment, Tehran, 2008.

14. T. Abbasi, S.A. Abbasi, The boiling liquid expanding vapour explosion (BLEVE) is fifty ... and lives on!, Journal of Loss Prevention in the Process Industries 21 (2008) 485–487.

15. T. Abbasi, S.A. Abbasi, The expertise and the practice of loss prevention in the Indian process industry: some pointers for the third world, Trans IChemE 83 (2005) 413–420.

16. F.I. Khan, S.A. Abbasi, Analytical simulation and PROFAT II: a new methodology and a computer automated tool for fault tree analysis in chemical process industries, Journal of Hazardous Materials 75 (2000) 1–27.

17. V.C. Marshall, Consequence analysis of explosion hazards, in: K.V. Raghavan, G. Swaminathan (Eds.), Hazard Assessment and Disaster Mitigation in Petroleum and Chemical Process Industries, Oxford & IBH, New Delhi, Calcutta, 1996, pp. 69–82.

18. FM Global: Factory Mutual Global, Property Loss Prevention Data Sheets 7-0, Factory Mutual Insurance Company, Boston, 2006.

19. R.W. King, J. Magid, Industrial Hazard and Safety Handbook, Butterworths, London, 1980.

20. L.A. Medard, Accidental Explosions, Wiley & Sons, New York, 1989.

21. G. Wells, Major Hazards and their Management, Institution of Chemical Engineers (IChemE), Great Britain, 1997.

22. CCPS, Guidelines for Evaluating the Characteristics of Vapor Cloud Explosions, Flash Fires, and BLEVEs, American Institute of Chemical Engineers, 1994.

23. R.J. Martin, A. Reza, L.W. Anderson, What is an explosion? A case history of an investigation for the insurance industry, Journal of Loss Prevention in the Process Industries 13 (2000) 491–497.

24. D.A. Crowl, J.F. Louvar, Chemical Process Safety Fundamentals with Applications, second ed., Prentice Hall, 2002.

25. [25] F.I. Khan, S.A. Abbasi, Accident hazard index: a multi-attribute method for process industry hazard rating, Process Safety and Environmental Protection 75 (1997) 217–224.

26. F.I. Khan, S.A. Abbasi, Risk analysis of a chloralkali industry situated in a populated area using the software package MAXCRED-II, Process Safety Progress 16 (1997) 172–184.

27. D.A. Crowl, Understanding Explosions. Centre for Chemical Process Safety, American Institute of Chemical Engineers, New York, 2003.

28. J.L. Woodward, Estimating the Flammable Mass of a Vapor Cloud, Center for Chemical Process Safety/AIChE, 1998.

29. R.K. Eckhoff, Explosion Hazards in the Process Industries, Gulf Publishing Company, Austin, TX, 2005.

30. R.P. Koopman, D.L. Ermak, Lessons learned from LNG safety research, Journal of Hazardous Materials 140 (2007) 412–428.

31. NFPA 921: Guide for Fire and Explosion Investigations, National Fire Protection Association, 2008.

32. F.I. Khan, S.A. Abbasi, Multivariate hazard identification and ranking system, Process Safety Progress 17 (1998) 157–170.

33. F.I. Khan, S.A. Abbasi, Inherently safer design based on rapid risk analysis, Journal of Loss Prevention in the Process Industries 11 (1998) 361–372.

34. H.S. Ledin, A Review of the State-of-the-Art in Gas Explosion Modeling, Health and Safety Laboratory Report HSL/2002/02, Crown copyright 2002.

35. A.A. Pekalski, J.F. Zevenbergen, S.M. Lemkowitz, H.J. Pasman, A review of explosion prevention and protection systems suitable as ultimate layer of protection in chemical process installations, Trans IChemE, Part B, Process Safety and Environmental Protection 83 (2005) 1–17.

36. W. Bartknecht, Explosions: Course, Prevention, Protection, Springer, Berlin, 1981.

37. H.A. Duxbury, A.J. Wilday, The design of reactor relief systems, Process Safety and Environmental Protection 68B (1990) 24–30.

38. H.G. Fisher, DIERS research program on emergency relief systems, Chemical Engineering and Processing 8 (1985) 33–42.

39. H.G. Fisher, An overview of emergency system design practice, Plant/ Operations Progress 10 (1991) 1–17.

40. J.E. Huff, Relief system design scope of CCPS effluent handling guidelines, Plant/Operations Progress 11 (1992) 53–71.

41. D.J. Lewis, Unconfined vapor clouds explosions—historical perspective and predictive method based on incident record, Progress in Energy and Combustion Science 6 (1980) 151.

42. F.I. Khan, S.A. Abbasi, Major accidents in process industries and an analysis of causes and consequences, Journal of Loss Prevention in the Process Industries 12 (1999) 361–378.

43. T. Abbasi, S.A. Abbasi, Pollution Control, Climate Change, and Industrial Disasters, Discovery Publishing House, New Delhi, 2009.

44. H. Itagaki, M. Tamura, http://shippai.jst.go.jp/en/Detail?fn=0&id=CC1000176&kw+ogaki&nkw=1971&knw=august, Last accessed 05 September 2009.

45. T. Dokter, Fire and explosion hazard of chlorine-containing systems, Journal of Hazardous Materials 10 (1985) 73–87.

46. http://ehstoday.com/fireemergencyresponse/news/csb-bayer-cause- 4109/Last Accessed September 5,2009.

47. A. Luketa-Haulin, A review of large-scale LNG spills: experiments andmodeling, Journal of Hazardous Materials A132 (2006) 119–140.

Dust Collector Explosions: A Quantitative Hazard Evaluation Method

Robert Zalosh

Firexplo, Wellesley, MA, USA

ABSTRACT

A methodology is presented to evaluate the explosion hazard of typical bag and cartridge dust collectors. The evaluation accounts for the expected development of suspended dust concentrations greater than the dust MEC during the normal pulsing of the bags or cartridges to remove part of the attached dust. Equations are presented to calculate these concentrations and also the associated partial volume explosion pressures resulting from the ignition of these dust clouds. Five quantitative examples are presented. The methodology also includes considerations of potential upset condition full volume explosions associated with the detachment of about half the dust on the bags or cartridges. A flow chart is offered to implement this hazard evaluation method for special situations in which the need for dust explosion protection may not be obvious.

INTRODUCTION

According to many dust explosion incident compilations, more dust explosion incidents have occurred in dust collectors than any other type of equipment (CSB, 2006, Going and Lombardo 2007, and Zalosh et al, 2005). Although the basic phenomena and typical explosion scenarios pertinent to the dust collector explosion hazard are well known, apparently there has not yet been a quantitative hazard evaluation methodology published.

NFPA 654 (2013) and other combustible dust standards require explosion protection for dust collectors when an explosion hazard exists in the collector. Paragraph 6.1.7 of NFPA 654 (2013) states that an explosion hazard exists "where both of the following conditions are possible:

- Combustible dust is present in sufficient quantity to cause enclosure rupture if suspended and ignited.
- A means of suspending the dust is present."

In order to implement the NFPA 654 (2013) explosion hazard criteria, it is necessary to estimate the amount of combustible dust in a collector during normal operation, and calculate the explosion pressure expected if all that dust were suspended and ignited. It is also necessary to evaluate various means of suspending the dust in the collector during both normal and abnormal operation.

The following section provides equations and test data to estimate how much dust can be expected to be attached to the filtration media in a collector during normal operation, and how that amount is influenced by the collector operating parameters. It also provides a methodology for estimating the amount and concentrations of dust dispersed during filter pulsing. Section 3 of the paper uses the amounts and concentrations of dispersed dust in the collector to estimate partial volume explosion pressures, and compares those pressures to typical strengths of collectors.

The fourth section of the paper describes possible scenarios in which significantly more than the normal amount of dust is dispersed within the collector such that an ignition would be more likely to produce a full volume dust explosion. The fifth section of the paper combines the results of the first four sections to present a quantitative dust explosion hazard evaluation method for filtration dust collectors.

DUST ACCUMULATIONS AND DUST CLOUDS IN FILTRATION COLLECTORS

Dust Collector Design and Operation

The most commonly used dust collector is the baghouse illustrated in Figure 1. Dust laden air flows into the baghouse where it encounters a number of vertical filter bags suspended below a tube sheet that separates the clean air from the dirty air. The porous filter bags capture the dust on their exterior surfaces while the air flows into the bags and then up through the clean air plenum above the tube sheet. An exhaust fan, not shown in Figure 1, located in the clean air duct pulls the air through the dust collection ducting and baghouse.

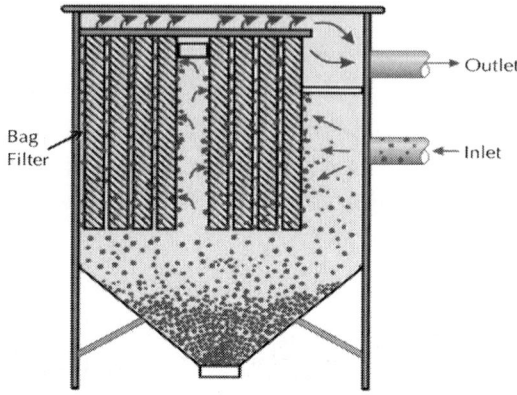

Figure 1: Baghouse schematic diagram.

Another widely used type of filter collector is the cartridge collector illustrated in Figure 2. Here the filter cartridges are usually horizontal and the dust laden air flows down and over the cartridges such that the dust is deposited on the cartridges while the dust free air flows through the cartridges on its way to the clean air outlet. As indicated in Figures 1 and 2, dust that is not captured by the bags or cartridges falls into a hopper at the bottom of the collector.

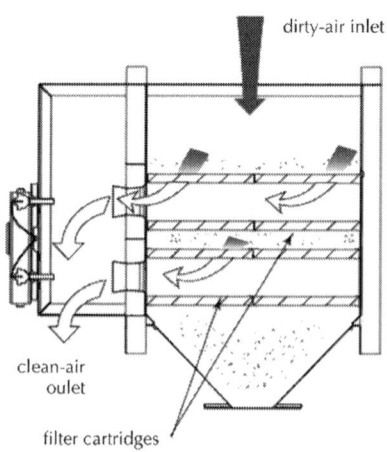

Figure 2: Typical cartridge collector diagram showing normal operation.

The accumulated dust on the external surfaces of the bags and cartridges forms a filter cake shown in the normal operation diagram in Figure 3. Periodically, a portion of the filter cake is blown off the bags or cartridges by a pulse of compressed air directed into the bag or cartridge interior as shown in the filter cleaning operation diagram in Figure 3. The compressed air pulse temporarily interrupts and reverses the normal air flow direction into the bag/cartridge.

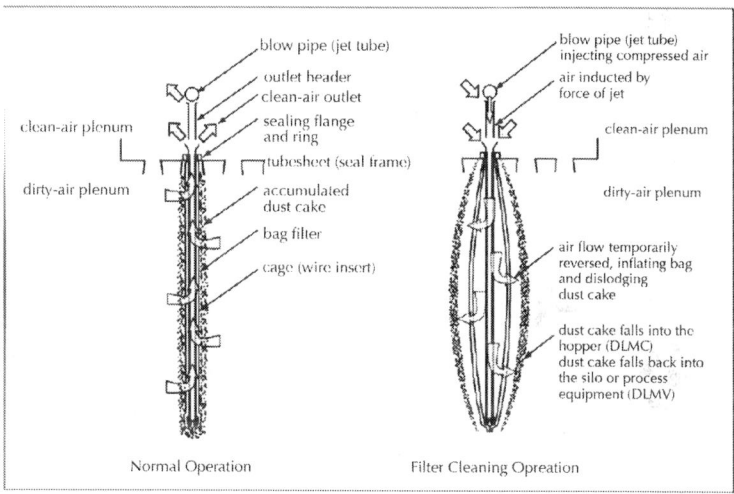

Figure 3: Dust cake formation and partial dispersal during pulsing.

Dust Loads on Bags and Cartridges

The dust filter cake load density (collected dust mass per unit bag surface area) and corresponding filter cake height depend on the inlet dust concentration, the filter media characteristics, the air pressure drop across the filter during dust collection, and the time interval or differential pressure that triggers air pulse reverse flow to initiate dust partial detachment from the bags. As an example, the filter cake height reported in (Saleem, 2011) for an inlet dust concentration of about 7 g/m^3 varied from 0.1 mm to 0.5 mm as the air pressure drop across the bags varied from 420 Pa to 1225 Pa (1.7 to 4.9 inches of water). Slightly lower heights in the range 0.075 to 0.4 mm were measured at inlet concentrations of 4.5 to 4.8 g/m³. More detailed measurements accounting for patches of filter cakes on the filters and a wide variation of pressures showed a distribution of cake heights with an upper bound of about 1 mm at high differential pressures of about 1500 Pa (6 in water). Even larger heights up to 3 mm have been measured with much larger dust inlet concentrations of about 200 g/m³ (Chen & Hsiau, 2009).

The relationship between filter cake load density, m", and filter cake height, h, is

$$m" = \rho_b h = \rho_p (1 - \epsilon)h$$

(1)

where ρ_b is the compacted bulk density of the filter cake just before pulsing, ρ_p is the dust particle density, and ϵ is the filter cake porosity. Filter cake porosities depend on the dust properties and the pressure drop (or corresponding air velocity) across the filter bag/cartridge. Chen & Hsiau, 2009 found ϵ to be in the range 0.65 to 0.71 for fly ash, whereas Sliva et al (1999) report lower porosities in the range 0.28 to 0.33 for a fine cohesive powder. Dust particle densities for a variety of combustible dusts are listed in Table 1, and the corresponding filter cake load densities are shown for a porosity of 0.55 and 0.70, and filter cake depths of 0.08 mm and 0.50 mm, i.e. the range cited above for inlet concentrations of 4.5 g/m³ to 7 g/m³.

Table 1: Filter cake load densities, m″ (g/m²), for various dust materials and cake depths

Material	Particle Density (g/cm³)	m″ for ε=0.55, h=0.08 mm	m″ for ε=0.55, h=0.50 mm	m″ for ε=0.70, h=0.08 mm	m″ for ε=0.70, h=0.50 mm
Aluminum	2.72	979	6120	653	4080
Carbon black	1.8	648	4050	432	2700
Coal (Pittsburgh)	1.37	493	3083	329	2055
Corn Starch	1.55	558	3488	372	2325
Milk fat powder	0.94	338	2115	226	1410
Polystyrene Expanded	~ 0.7	252	1575	168	1050
Sawdust	1.06	382	2385	254	1590
Sugar	1.60	576	3600	384	2400
Titanium	4.51	1624	10148	1082	6765

imply that there would be a different value for K for each row, such that the variation from minimum to maximum K would be approximately a factor of 3. This implies that the maximum K might be in the range 1.5 to 2.0.

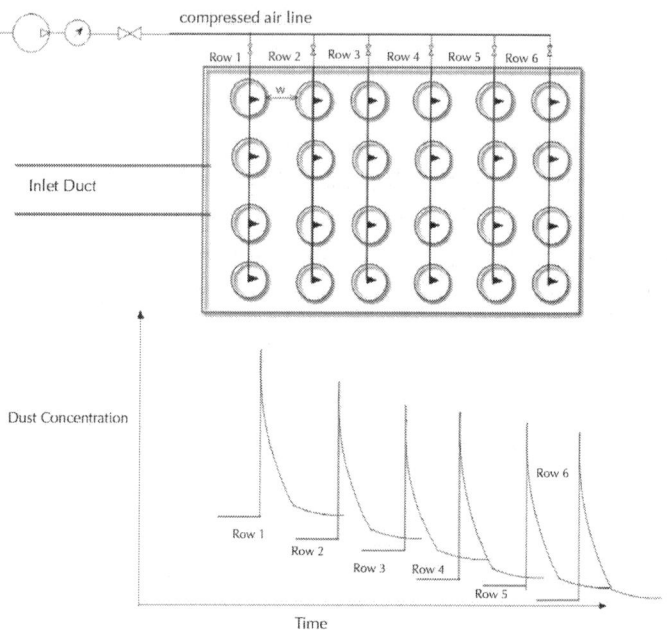

Figure 4: Dust concentrations during pulsing of bag rows.

Since the dust removed from each bag row during pulsing will fill the space between adjacent rows, the transient concentration during pulsing, c_{pulse}, is

$$c_{pulse} = \frac{\left(N_{bag}/N_{row}\right)m_{dbag}}{2wLH_{bag}}$$

(4)

where N_{row} is the number of bag rows, w is the distance between rows, L is the length of each row, and H_{bag} is the length of each bag or cartridge. Using Equation 3 for $N_{bag}m_{dbag}$, the maximum dust concentration during steady-state pulsing is

$$c_{pulse} = \frac{c_{in} Q_a t_{cycle} K_{max}}{2 w L H_{bag} N_{row}}$$

(5)

As an example, consider the baghouses listed in Table 2 used to collect sawdust with an inlet concentration of 5 g/m^3. The other pertinent parameters needed in Equation 5 are listed in Table 3 along with the calculated values of c_{pulse}. All five calculated c_{pulse} values are significantly larger than the 30 g/m^3 to 60 g/m^3 Minimum Explosible Concentrations (MECs) for wood dusts listed in Table A.1of Eckhoff (2003). Therefore, the normal pulsing of the bags or cartridges under these conditions produces explosible concentrations even though the inlet dust concentration is an order-of-magnitude smaller than the MEC concentration.

Table 3: Calculated Dust Pulse Concentrations for 5 g/m^3 inlet concentration

Collector #	Q_a (m^3/s)	t_{cycle} (s)	N_{row}	w (m)	H_{bag} (m)	c_{pulse} (g/m^3)
1	2.36	242	20	0.071	3.39	274
2	0.94	120	4	0.170	1.46	311
3	3.28	70	10	0.152	2.44	156
4	1.42	48	4	0.168	1.42	354
5	4.72	194	16	0.168	1.42	589

In the case where the reverse air pulse is triggered when the pressure drop, ΔP, across the dust laden bags or cartridges reaches or exceeds a specified value, ΔP_{pulse}, the following relationship between ΔP and the filter cake mass load density (Silva et al, 1999, and Chen & Hsiau, 2009) can be invoked.

$$\Delta P = \Delta P_{Fabric} + \Delta P_{Cake} = K_F u + K_C u \, m''$$

(6)

where ΔP_{Fabric} is the pressure drop across the filter media, ΔP_{Cake} is the pressure drop across the filter cake, u is the superficial air velocity through the cake and filer media, and K_F and K_C are resistance coefficients that are developed in the cited references and in Calle et al, 2002 in terms of the theory of flow through a porous medium composed of fibers or particulates. They are directly proportional to the filter media thickness and filter cake thickness, and depend also on porosity and vary inversely as the square of the particle/fiber size. Values of K_F obtained from the cited references and known pressure drops through clean media are in the range 1.2×10^3 Pa-s/m to 1.5×10^4 Pa-s/m. Values of K_C obtained from the cited references for filters made of various polymer fibers and noncombustible dust particle sizes in the range 5 µm to 20 µm are in the range 1.2×10^5 1/s to 2.3×10^5 1/s. Calculations using representative values of K_F and K_C in this range have been used in the following equation derived from Equation 6 to calculate the values of m" corresponding to the pulse onset pressure drop ΔP_{pulse}.

$$m"_{pulse} = \frac{\Delta P_{pulse} - K_F u}{K_C u}$$

(7)

The calculated values of m"$_{pulse}$ for a typical value of $\Delta P_{pulse} = 750$ Pa and for the values of u calculated for the five example collectors listed in Table 1 range from about 300 g/m² to about 1600 g/m². The average dust concentration during pulsing into the collector dirty side volume, V_{dirty}, is given now by

$$c_{pulse} = \frac{(m"_{pulse} - m"_{att}) A_{bag} N_{bag}}{V_{dirty}}$$

(8)

where m"$_{att}$ is the value of m" that remains attached to the bags during pulsing. Using a value of m"$_{att}$ of 200 g/m², values of c_{pulse} have been calculated from Equation 8 for the five example dust collectors. In the case of the three cartridge dust collectors, these values ranged from 930 g/m³ to 19000 g/m³, and even larger values were obtained for the two example baghouses. All these values of c_{pulse} are much larger than the corresponding values shown in Table 3 for timed periodic pulsing and an inlet concentration of 5 g/m³. This comparison shows that filter pulsing based on a typical maximum allowable pressure drop across the filters produces much higher dust concentrations in the collector than the timed periodic filter reverse air pulsing.

PARTIAL VOLUME EXPLOSION HAZARD DURING FILTER PULSING

If a competent ignition source is located in the volume occupied by the pulsed bag dust cloud, and the dust cloud concentration is above the MEC, a partial volume dust explosion will occur. The pressure developed in this partial volume explosion within the dust collector will depend on the dust concentration, the collector partial volume fraction, x_{pv}, occupied by the explosible dust cloud, the combustible dust closed vessel explosion pressure at that dust concentration, the level of turbulence at ignition, and the complicating effects of the collector bags/cartridges and additional dust that can be dispersed by the air flow associated with the developing explosion. An approximate calculation of the partial volume explosion pressure, P_{pv}, for this scenario can be developed by neglecting the effects of the bags/cartridges, additional dust in the collector, and venting through the inlet duct, and using the following partial volume explosion pressure equation originally derived for gas explosions (Jo and Park, 2004).

$$\frac{P_{pv}}{P_a} = \left[1 - x_{pv} + x_{pv}\left(\frac{P_{max}}{P_a}\right)^{1/\gamma}\right]^{\gamma}$$

(9)

where x_{pv} is the ratio of the combustible cloud volume to the enclosure volume, i.e. the partial volume fraction, P_{max} is the closed vessel enclosure pressure for that combustible mixture, P_a is atmospheric pressure, and γ is the ratio of specific heats (for air = 1.4). In the dust collector pulsing partial volume explosion scenario,

$$x_{pv} = \frac{2wLH_{bag}}{V_{dc}}$$

(10)

where V_{dc} is the dust collector dirty side volume, including the volume below the bags.

Since the calculated values of c_{pulse} shown in Table 3 are at least three times the MEC, it is reasonable for the sake of simplification to take P_{max} to be the P_{max} value for the dust in the collector, as determined by standard tests such as ISO 6184-1. These values for wood dust are in the range 6 to 9 (Eckhoff, 2003).Figure 5 is a plot of P_{pv} versus x_{pv} using Equation 9 with P_{max}/P_a values of 6 and 9.

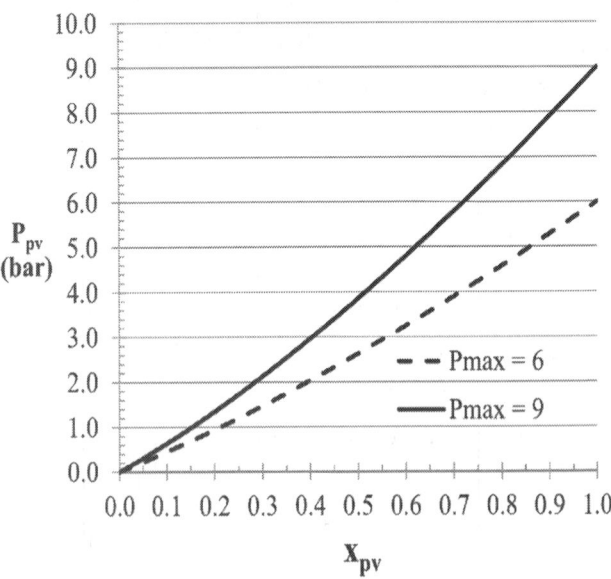

Figure 5: Partial volume explosion pressure versus partial volume fraction.

Calculated values of x_{pv} using Equation 10 for the five example dust collectors are shown in Table 4 along with the corresponding values of P_{pv} for P_{max} values of 6 and of 9. The values of V_{dc} shown in Table 2 did include the hopper in some cases whereas the hopper volume was considered negligible compared to the large rectangular main body of the collector in other cases. The minimum value of P_{pv} is 0.19 bar, and the maximum value is 0.97 bar. The prismatic geometry and construction of these five collectors suggests that the threshold pressure for permanent damage would be in the range 0.2 bar to 0.3 bar, and the structures would probably rupture when pressures exceed about 0.5 bar. Therefore, the bag pulsing partial volume sawdust explosion scenario for these five collectors would probably result in permanent damage to three of the collectors and rupture of the other two collectors, unless there was substantial reverse flow explosion venting through the inlet ducts. The problem with reverse flow venting through the inlet ducts is that the dust explosion could potentially propagate to upstream equipment placing additional personnel and structures in jeopardy. Clearly, some form of dust explosion protection is needed for all five example collectors for the hypothesized sawdust collection conditions.

Table 4: Dust Collector Explosion Pressures for Bag Pulsing Scenario

Collector #	x_{pv}	P_{pv} (bar g) for P_{max} = 6 bar g	P_{pv} (bar g) for P_{max} = 9 bar g
1	0.059	0.26	0.36
2	0.149	0.68	0.97
3	0.061	0.27	0.37
4	0.124	0.56	0.80
5	0.043	0.19	0.26

DUST DISPERSAL AND EXPLOSION DURING UPSET SCENARIOS

A comprehensive hazard analysis for filter type dust collectors should account for abnormal conditions potentially leading to larger and more concentrated clouds of dust than the normal bag pulsing scenario described above. One category of these abnormal or upset condition scenarios would be abnormally high inlet dust concentrations because of some unusual dust generation condition. An example is the sudden extrication of a blockage in the collector inlet line ducting such that the accumulated dust in the inlet duct enters the collector as the blockage is removed. The Chemical Safety Board (2005) report of the CTA Acoustics explosion describes a collector duct blockage removal being the initiating event that led to a fire and an eventual propagating explosion with devastating results.

Another category of upset scenarios would be more than the usual amount of dust being detached from the bags. Some examples of possible upset conditions leading to excess dust being detached from bags/cartridges are the following:

- A temporary interruption in the compressed air supply for pulsing, allowing much more than the normal amount of dust to be attached to the bags. Resumption of pulsing would then detach more dust than normal.
- Shaking of the bags/cartridges due to maintenance or repair activities such as the removal of bags for replacement.
- Discharge of a fire extinguisher towards the bags in response to a smoldering fire in the collector.
- Discharge of compressed air from a dry pipe sprinkler line in the collector as the sprinkler head opens in response to a small fire in the collector.

- An external event such as an explosion, seismic activity, or impact perhaps due to an industrial truck colliding with the collector.

Most, if not all, of the preceding examples have occurred in actual incidents. One such Example 2 excess dust detachment during repairs type occurrence involved the erection of scaffolding in a large dust collector hopper without first removing the dust laden bags above the hopper. Assembling the scaffolding entailed banging on the scaffold pipes and shaking the collector structure. The dust rained down into the hopper and was ignited by a halogen lamp used by the workers erecting the scaffolding. The resulting explosion seriously burned eight workers located in and around the hopper.

Koch et al (1996) have compiled test data on the removal of dust accumulations from filters due to sudden accelerations of the filters. This data is pertinent to the shaking of the bags and cartridges due to maintenance/repairs (Example 2) and to external events (Example 5). The acceleration required to remove 50% of the attached dust was in the range 100 m/s^2 to 200 m/s^2 over a wide range of dust loadings in three separate test programs. Since many older dust collectors used mechanical shaking/thumping as the means to detach filter cakes, it is reasonable to assume that as much as 50% of the filter cake mass can be detached by repeated banging on the collector, which might occur when personnel attempt to break up the bridging of dust in the collector hopper.

Detachment of about 50% of the filter cake from the bags or cartridge due to deliberate or accident impact would, for the example collectors at light loads, suspend 15 kg to 185 kg. Since the dirty side volumes of these dust collectors range from 4 m^3 to 24 m^3, it is clear that such detachments would produce suspended dust concentrations far above 1000 g/m^3 even if the detachment occurred over several seconds. These concentrations easily reach or exceed the concentrations associated with measured P$_{max}$ values in dust explosibility testing. Therefore, an explosion resulting from many of the upset scenarios would result in pressures far above the pressures computed above for partial volume deflagrations.

SUGGESTED EXPLOSION HAZARD METHODOLOGY WITH EXAMPLE

The discussion above and the numerous dust collector explosion incidents demonstrate that the NFPA 654 (2013) explosion hazard criteria are satisfied for most filter media collectors handling combustible dusts. Thus, the default practice should be to provide explosion protection for these collectors assuming that they cannot withstand explosion pressures corresponding to the P_{max} values of the combustible dusts. Explosion protection options for dust collectors are described in many combustible dust standards such asNFPA 654 (2013) and FM Global Data Sheets 7-73 and 7-76, and in papers such as Going and Lombardi (2007). Explosion venting, which is the most common type of explosion protection for baghouses, needs to account for bag/cage displacement and possible partial obstruction of vent openings. Vent design methods and vent locations to account for these phenomena are described in NFPA 68.

There are special situations in which a collector-specific quantitative explosion hazard evaluation may be warranted. These situations include: 1) collectors for marginally explosible dusts with P_{max} values less than about 5 bar, 2) small dust collectors (NFPA 654 (2013) specifies a dirty side volume of smaller than 0.2 m^3 as the criterion for small), 3) large dust collectors located outdoors, 4) dust collectors with inlet duct dust concentrations less than about 1% of the MEC, 5) collectors with a pressure resistance on the order of 1 bar or larger, 6) low K_{ST} dusts producing relatively slow deflagration pressure increases such that venting through the inlet duct may effectively prevent damage to the collector and upstream equipment, and 7) adoption of the Performance Based Design Option in NFPA 654 (2013) and other NFPA combustible dust standards. The suggested explosion hazard methodology for these situations is shown in Figure 6 flow chart.

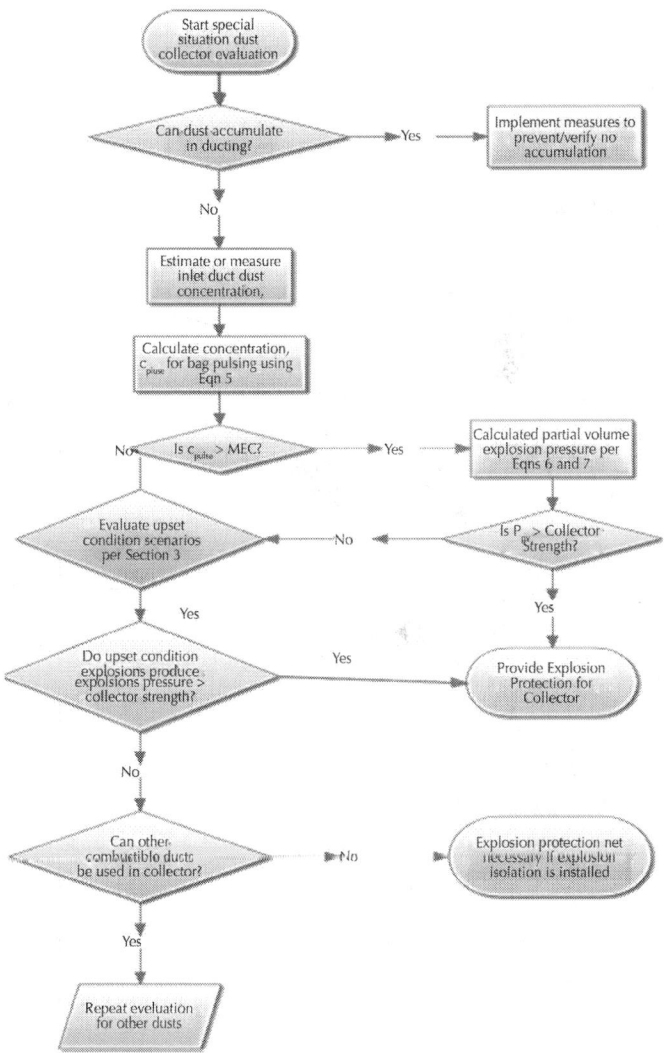

Figure 6: Suggested special situation explosion hazard evaluation flow chart.

As an example in the use of the flow chart, consider a situation involving the use of dust collector # 3 for collecting wood dust with an inlet dust concentration less than about 0.5 g/m³, i.e. less than about 1% of the MEC for many types of wood dust including the dust in the facility being evaluated. Since Equation 5indicates that c_{pulse} is linearly proportional to c_{in}, and since c_{pulse} for a c_{in} of 5 g/m^3 was 156 g/m^3 per Table 3, the corresponding c_{pulse} for this special situation would be less than 15.6 g/m^3, i.e. below the dust MEC. Following the flow chart for this example, we are now at the upset condition evaluation. In this case, at least some of the Section 4 upset conditions are credible, so we need to evaluate the expected explosion pressure for a full volume dust explosion. The total amount of dust in the collector for the 200 g/m^2 light load density would be 47 kg. The dirty side volume dust concentration corresponding to half the 47 kg being dispersed in this collector would be about 970 g/m^3. This concentration is within the range of worst-case concentrations that produce the wood dust P_{max} full volume explosion pressure of 6 to 9 bar g. This collector strength is only on the order of 0.5 bar, so explosion protection is needed in this special situation.

There can be applications in which the filter cake load density is even lower than the 200 g/m^2 light load density used in the preceding calculations. Filter cake load densities corresponding to the Dust Holding Capacities measured in standardized tests for general ventilation air filters, such as ASHRAE 52.2 (2007)and EN 779 (2012), are lower in many cases, but the applicability of this data to baghouses and other industrial dust collectors is problematic. If and when similar consensus standard test methods are developed and implemented for industrial dust collection applications with pulsed filter dust detachment, the data should be useful for implementation of the dust collection hazard evaluation method described here. In the meantime, practitioners can use either data in Table 1 or estimates based on Equation 1 along with informed estimates of filter cake height, h, for the specific application being evaluated.

CONCLUSIONS

A methodology has been formulated to estimate the dust concentrations during the pulsing of filter media dust collectors to remove part of the dust attached to the bags or cartridges for a known steady-state inlet dust concentration. In the five examples of various size bag and cartridge collectors, these concentrations are well above the sawdust MEC. The partial volume sawdust explosion pressures for these five collectors for the postulated inlet dust concentration are in the range 0.2 bar to 1 bar. These explosion pressures would be expected to produce significant damage and possible rupture of the five example collectors.

A methodology has also been developed to evaluate the potential for full volume dust collector explosions due to various upset conditions. This methodology suggests that most, if not all, of the upset conditions would result in explosion pressures approaching or approximately equal to the P_{max} value for a given combustible dust. Hence explosion protection is warranted for most filter media dust collectors handling combustible dusts.

There are several special situations in which the need for explosion protection is less obvious, and a quantitative explosion hazard evaluation would be useful. A flow chart to conduct such special situation hazard evaluations has been developed using the described evaluations for partial volume deflagrations associated with pulsing, and for possible full volume deflagrations associated with upset conditions.

REFERENCES

1. ASHRAE Standard 52.2-2007 Addendum b (2008) "Method of Testing General Ventilation Air-Cleaning Devices for Removal Efficiency by Particle Size," American Society of Heating, Refrigerating and Air-Conditioning Engineers, Inc.

2. Callé, S., Contal, P., Thomas, D., Bémer, D, and Leclerc, D., (2002) "Description of the clogging and cleaning cycles of filter media," Powder Technology, 123, pp 40-52.

3. Chemical Safety Board, (2006) "Combustible Dust Hazard Study," Chemical Safety Board Report No. 2006-H-1.

4.   Chemical Safety Board, (2005) "Investigation Report: Combustible Dust Fire and Explosions (7 Killed, 37 Injured)," Report No. 2003-09-I-KY.

5.   Chen, Y-S. and Hsiau, S-S., (2009) "Influence of filtration superficial velocity on cake compression and cake formation," Chemical Engineering and Processing: Process Intensification, v 48, pp 988-996.

6.   Eckhoff, R.K. (2003). Dust explosions in the process industries. Gulf Professional Publishing, Amsterdam, third edition.

7.   EN 779 (2012) "Particulate air filters for general ventilation. Determination of the filtration performance."

8.   FM Global (2009) "Dust Collectors and Collection Systems," Loss Prevention Data Sheet 7- 73.

9.   FM Global (2013) "Prevention and Mitigation of Combustible Dust Explosion and Fire," Loss Prevention Data Sheet 7-76.

10.   Going, J. and Lombardo. T. (2007) "Dust Collector Explosion Prevention and Control," Process Safety Progress, v 26, pp 164-176.

11.   ISO 6184-1 (1985) "Explosion protection systems -- Part 1: Determination of explosion indices of combustible dusts in air,"

12.   Jo, Y-D and Park, K-S (2004). "Minimum Amount of Flammable Gas for Explosion within a Confined Space," Process Safety Progress,v 23, pp 321-329.

13.   Koch, D., Seville, J., and Clift, R. (1996) "Dust cake detachment from gas filters," Powder Technology, v 86, pp 21-29.

14.   NFPA 68 (2013) "Standard on Explosion Protection by Deflagration Venting," National Fire Protection Association.

15.   NFPA 654 (2013). Standard for the Prevention of Fire and Dust Explosions from the Manufacturing, Processing, and Handling of Combustible Particulate Solids. National Fire Protection Association.

16.   Saleem, M., Khan, R., Tahir, M., and Krammer, G, (2011) "Experimental study of cake formation on heat treated and

membrane coated needle felts in a pilot scale pulse jet bag filter using optical in-situ cake height measurement," Powder Technology, v 214, pp 388- 399.

17.   Simon, X., Bémer, D., Chazelet, S., Thomas, D., and Régnier, R., (2010) "Consequences of high transitory airflows generated by segmented pulse-jet cleaning of dust collector filter bags," Powder Technology, v 201, pp 37-48.

18.   Silva, C, Negrini, V., Aguiar, M, and Coury, J., (1999) Influence of gas velocity on cake formation and detachment, Powder Technology, v 101, pp 165-172.

19.   Zalosh, R., Grossel, S., Kahn, R. and Sliva, D. (2005) "Safely Handle Powdered Solids," Chemical Engineering Progress, December 2005, pp 23-30.

Chapter 3

Safety in the Mining Industry and the Unfinished Legacy of Mining Accidents: Safety Levers and Defense-in-depth for Addressing Mining Hazards

Joseph H. Saleh and Amy M. Cummings

School of Aerospace Engineering, Georgia Institute of Technology, Atlanta, GA 30332, USA

ABSTRACT

Mining remains one of the most hazardous occupations worldwide and underground coal mines are especially notorious for their high accident rates. In this work, we provide an overview of the broad and multi-faceted topic of safety in the mining industry. After reviewing some statistics of mining accidents in the United States, we focus on one pervasive and deadly failure mode in mines, namely explosions.

The repeated occurrence of mine explosions, often in similar manner, is the loud unfinished legacy of mining accidents and their occurrence in the 21st century is inexcusable and should constitute a strong call for action for all stakeholders in this industry to settle this problem. We analyze one such recent mine disaster in which deficiencies in various safety barriers failed to prevent the accident initiating event from occurring, then subsequent lines of defense failed to block this accident scenario from unfolding and to mitigate its consequences. We identify the technical, organizational, and regulatory deficiencies that failed to prevent the escalation of the mine hazards into an accident, and the accident into a "disaster". This case study provides an opportunity to illustrate several concepts that help describe the phenomenology of accidents, such as *initiating events*, *precursor* or *lead indicator*, and *accident pathogen*.

Next, we introduce the safety principle of defense-in-depth, which is the basis for regulations and risk-informed decisions by the US Nuclear Regulatory Commission, and we examine its relevance and applicability to the *mining system* in support of accident prevention and coordinating actions on all the safety levers, technical, organizational, and regulatory to improve mining safety. The *mining system* includes the physical confines and characteristics of the mine, the equipment in the mine, the individuals and the organization that operate the mine, as well as the processes and regulatory constraints under which the mine operates. We conclude this article with the proposition for the establishment of defense-in-depth as the guiding safety principle for the mining industry and we indicate possible benefits for adopting this structured hazard-centric system approach to mining safety.

INTRODUCTION

Mining remains one of the most hazardous occupations worldwide, and underground coal mines are especially notorious for their high accident rates. Recent media coverage of major fatal accidents in Ukraine, China, South Africa, and the United States highlight a growing public awareness of the dangers of mining. Unfortunately these reported events constitute only the tip of the "safety iceberg" in an industry that remains significantly important in many parts of the world and constitutes an essential economic activity for many communities.

In mining as well as in other hazardous industries, various safety levers can be acted upon to modify the exposure to the inherent risks involved, to minimize their likelihood of occurrence, and to contain or mitigate the consequences of accidents should they unfold. Decision-makers can adopt different attitudes and choices with respect to the conditions in these industries. Consider the following episode in which the comments of the protagonists are particularly revealing of different "archetype" attitudes typically adopted in the face of hazardous conditions. In November 2007, a major explosion occurred in a mine in Ukraine killing 90 miners. The Prime Minister at the time, Viktor Yanukovych was reported as saying, "this accident has proven once again [*emphasis added*] that a human is powerless before nature" (BBC News, 2007a and BBC News, 2007b). The Ukrainian President Viktor Yushchenko adopted a different attitude and charged that "the government had made insufficient efforts to re-organize the mining sector, particularly the implementation of safe mining practices." One finds in these comments on the one hand a resigned attitude to accept the inherent hazards of the industry (in this particular case it was a build-up of methane), on the other hand, a recognition that some safety levers exist, and that they can be acted upon to modify the exposure to the risks of mining (in this case, improved government oversight and regulation, and restructuring of the industry).

These two attitudes reflect typical perspectives on accident causation and system safety, and they are not unfamiliar to the

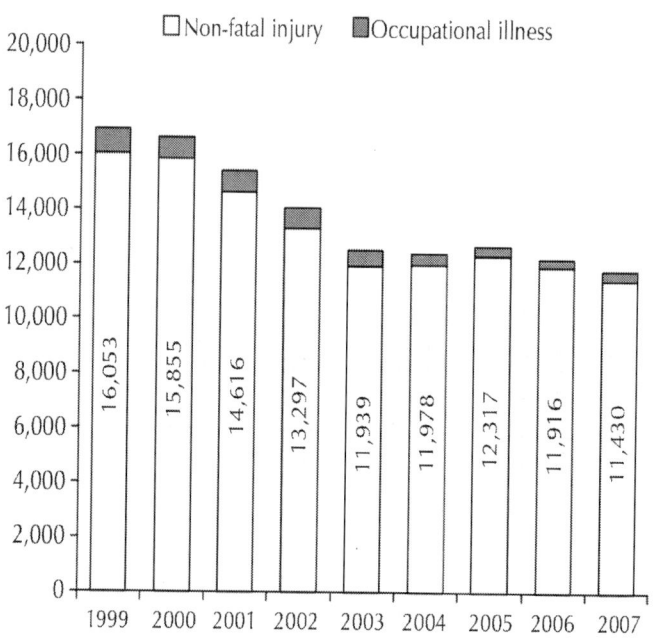

Figure 2: All US mining injuries and illnesses. *Source*: US Mining Safety and Health Administration

It is informative to look at broader historical trends in the evolution of fatalities and fatality rates in the mining industry. Fig. 3a shows the evolution of the number of fatalities in the US mining industry a century ago, between 1909 and 1918. The industry back then claimed the lives of a sobering 2600 miners on average every year, probably a result of a deadly combination of technical ignorance (of hazardous conditions, especially methane build-up and explosions), organizational recklessness, and lack of government oversight.

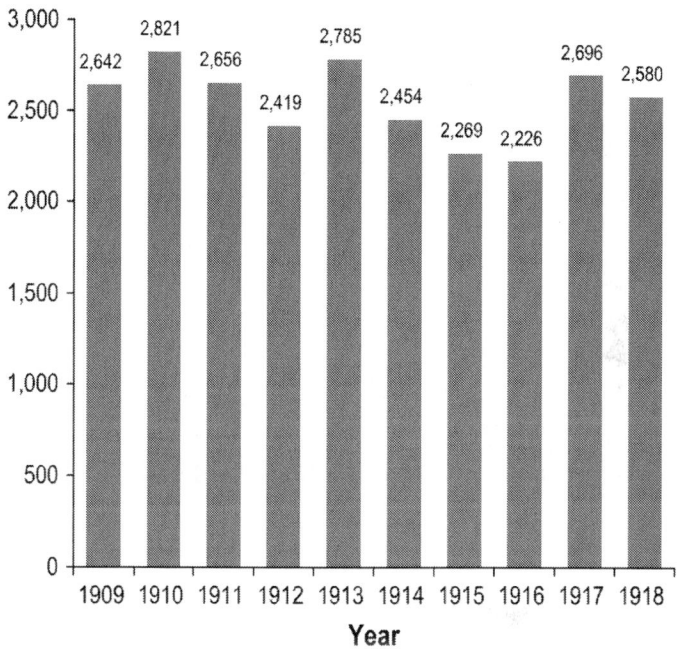

Figure 3a: US coal mining fatalities 1909–1918. *Source*: US Mining Safety and Health Administration.

Fifty years later, between 1959 and 1968, the US coal mining fatalities, shown in Fig. 3b on the same scale as that in Fig. 3a for comparative purposes, show a dramatic decrease from the earlier levels, to a yearly average of roughly 270 fatalities per year. The total number of workers in the US coal mines decreased roughly by half between these two periods (see Fig. 4b), therefore the order of magnitude decrease in the number of fatalities between 1909–1918 and 1959–1968 reflects a fivefold decrease in the fatality rate, a smaller (than the absolute value decrease) but significant safety improvement nonetheless.

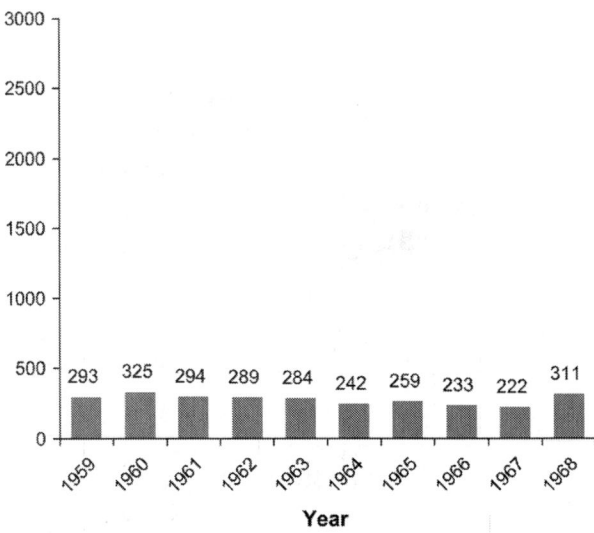

Figure 3b: US coal mining fatalities 1959–1968. *Source*: US Mining Safety and Health Administration.

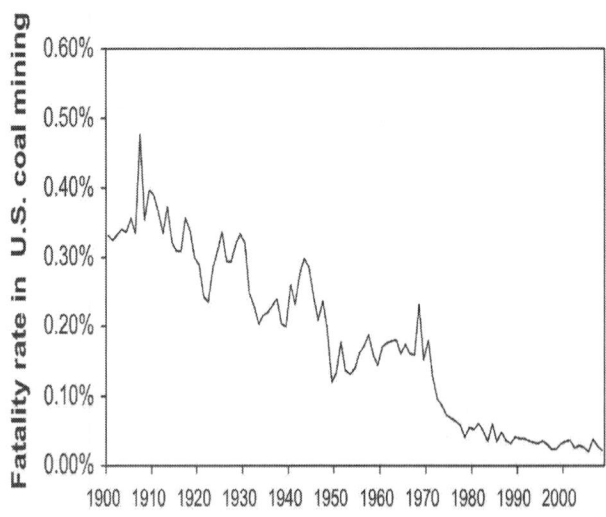

Figure 4b: Number of workers in the US coal mining. Office workers included starting in 1973. *Source*: US Mining Safety and Health Administration (MSHA).

This discussion invites a reconsideration of the statement by the Ukrainian Prime Minister quoted earlier, that "this accident has proven once again that a human is powerless before nature." Fig. 3a and Fig. 3billustrate that people are neither powerless nor omnipotent before nature, that safety levers do exist and can be acted upon to modify the exposure to the inherent risks in the mining industry, and thus in the long run improve the industry's safety record.

Since employment and fatalities data in the US coal mining exist dating back for over a century, fatality rates can be computed and their evolution over a long period of time displayed for (qualitative) visual assessment. Fig. 4a shows the evolution of the US coal mining fatality rate between 1900 and 2008. Displayed next to the fatality rate graph is the evolution of the number of workers in the US coal mines over the same period (Fig. 4b).

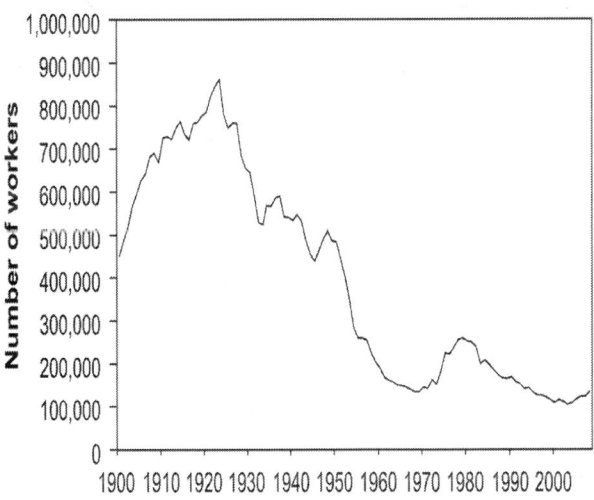

Figure 4a: Evolution of the fatality rate in the US coal mining between 1900 and 2008. *Source*: US Mining Safety and Health Administration (MSHA).

Fig. 4a shows several important trends in the evolution of the fatality rate. Three periods can roughly be distinguished as follows:

- *1900 – Late 1940s*: In this period, a significantly high fatality rate plagued the industry starting with roughly 0.4% a century ago or 400 fatalities per 100,000 miners. A slow downward trend is visible inFig. 4a during this period, with the fatality rate dropping from 0.4% in 1910 to roughly 0.2% in the late 1940s. It is worth pointing out that the US Bureau of Mines (USBM) was established in 1910 with the task of reducing mining accidents and improving conditions under which mining operations are conducted. The USBM is identified as having played a significant role in reducing these fatality rates (National Research Council, 2007), and the downward trend in the fatality rate during this period is indicative that careful government involvement and regulatory efforts can help make a difference in the safety record of a hazardous industry. Another important trend during this time period is the significantly high variability of the fatality rate, in sharp contrast for example with the variability of the fatality rate during the 1990s. This high variability is the result of major catastrophic events or mining disasters, each of which claimed the lives of a significantly large number of miners (see Fig. 5). For example, the highest peak in the fatality rate occurs in 1907 (towering at 0.48%) is the "signature" of the worst mining disaster in the US, the Monongah, West Virginia disaster in which 362 miners lost their lives due to a methane and coal dust explosion (it is believed though that the actual death toll was in excess of 500 [2] (Brigg, 1964).

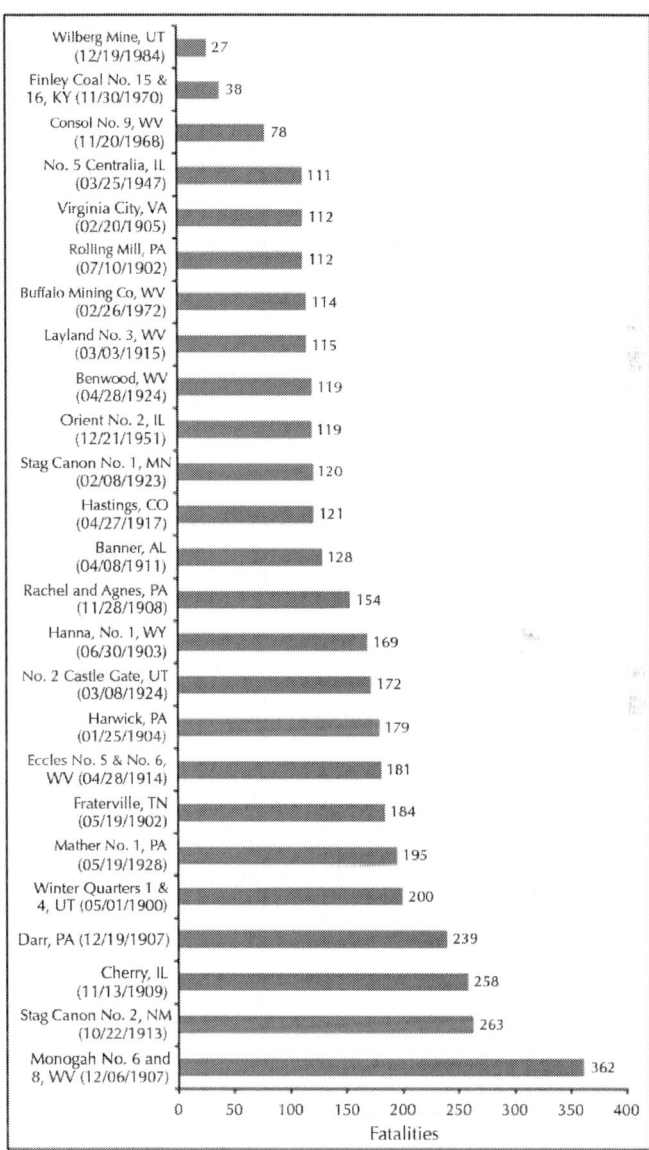

Figure 5: Sample of US coal mine disasters in the 20th century. *Source:* US National Institute of Occupational Health and Safety.

- *Late 1940s – early 1970s*: In this period, the fatality rate exhibits some variability around an average of 0.16% but it shows no distinctive upward or downward trend. The sharp peak in 1968 (at 0.23%) is the result of the worst mining accident during this period, namely the Farmington's Consol 9 (West Virginia) mine disaster in which 78 miners were killed. Following the Consol 9 disaster, Congress passed the Federal Coal Mine Health and Safety Act of 1969, a "comprehensive and more stringent than any previous Federal legislation governing the mining industry." This Coal Act of 1969, as it is known for short, required for example four annual inspections of every underground coal mine, and it significantly increased federal enforcement powers in coal mines, including the "establishment of criminal penalties for knowing and willful violations of safety standards." (History of Mine Safety and Health Legislation, undated). The effects of this legislation were to be seen in the third period discussed next.

- *Early 1970s – present*: This period exhibits both a low fatality rate around 0.03% and limited variability (compared with the previous two periods). This period starts with a sharp decrease in the fatality rate in the early 1970s. These observations indicate that on the one hand effective intervention has indeed paid off and compressed the fatality rate, and on the other hand, that major disasters with significantly large number of fatalities – which translate into variability of the fatality rate – have been if not eliminated at least kept at bay during this recent period. The small peak in the fatality rate in the lower right corner of Fig. 4a occurred in 2006 as a result of the Sago mine (West Virginia) disaster in which 12 miners were killed. For reference, over the last decade, the employment in the US mining industry hovered around 350,000 people.

Fig. 5 shows the deadliest coal mine disasters in the US in the 20th century. The data is displayed by increasing number of fatalities and is extensive starting from the No. 5 Centralia disaster in 1947 in which 111 miners were killed. The first three disasters however are only a small sample of more recent mining disasters, and many other accidents occurred that had between 111 and 38 fatalities but are not shown in this chart. Incidentally, the Mining Program within the US National Institute of Occupational Health and Safety defines a mine "disaster" as an event that involves five or more fatalities.

An extensive list of coal mining disasters worldwide can be found in Hugenard et al. (1996).

Although beyond the scope of the present work, it is sobering to reflect, beyond the statistics displayed inFig. 5, on the human tragedies behind these numbers and the traumatic consequences and scars that these events must have left in many communities and regions throughout the country.

The causes of mining accidents include explosions (methane, coal dust, or other), fires, rock and roof falls, landslides, blackdamp and toxic gases outbursts, and water inrush/sudden inundations (see Appendix Afor a brief discussion of such accidents). The majority of the mining disasters shown in Fig. 5 were the result of (methane + coal dust) explosions, as were most of recent large-scale mining disasters in China.[3]Accidents in the mining industry that result in a small number of fatalities typically include electrical equipment and machinery, powered haulage, and falls (or "slip, trip, and fall").

HEALTH AND SAFETY ISSUES IN THE MINING INDUSTRY

It is common in the mining industry to separately identify health issues and safety issues. This separate classification does not necessarily reflect mutually independent hazards, but it helps recognize a difference in the time scales of effect of the hazard sources. For example an explosion or a mine collapse will result in immediate traumatic injuries or fatalities, whereas a prolonged exposure to coal or silica dust can result in debilitating and fatal consequences for miners over the years in the form of lung diseases (e.g., black lung or Coal Workers Pneumoconiosis, CWP). Fig. 6 illustrates this safety versus health classification.

Initiating event: The initiating event of this disaster, the roof fall, gave repeated signs of its impending occurrence, but the seriousness of the situation – its increasing hazardous potential – was not appreciated.*Precursors* or *lead indicators* of the roof fall included the following:

- The roof at the accident location had been deteriorating 2 days before the disaster, and several miners heard at different occasions noises, popping sounds, and loud thumps "like pins breaking".

- In addition to these *acoustic* signatures of a hazardous condition, water was seen dripping from the roof. Although at first the amount was considered normal, the dripping did not abate for 2 days. Instead the dripping turned into pouring, a condition described in one accident report as "raining water" during the day of the disaster.

- Furthermore, cracks were seen in the roof, and over the 2 days prior to the fall, the cracks were noted to have increased in size.

- One more indication that the "top was working" (roof movement) was that a close-by cement brattice was "taking weight" and cracking, and some ribs were sloughing and rolling, a condition that worsened roughly an hour prior to the roof fall.

First accident pathogen [7] and flawed operational decision: These lead indicators of the initiating event, the roof fall, were not isolated events but came in support of a known adverse roof condition, a geological fault or "discontinuity" near the location of the accident. Unfortunately, instead of extra precaution in dealing with this known and deteriorating roof condition, a surprising operational decision was made, reflecting an improper safety culture at the mine and contributing to the development of the accident sequence: a large battery (weighing 6 tons) and battery charger were brought and placed under the frail roof.

On the afternoon of September 23, 2001, at 5:17 pm the roof fell and damaged the battery. The resulting methane outburst from the fall, along with the general "gassiness" of the mine created an explosive mixture in this particular section of the mine, an explosive mixture that was triggered at 5:20 pm by arcing from the battery recently brought to that location (see Fig. 7).

Intermediate event and consequences: The first explosion thus resulted from a confluence of factors, the immediate ones being the fact that the mine was "very gassy", that the roof condition was bad and deteriorating, and that an energized battery was brought at this particular location. Four miners were injured by this first explosion, three survived the disaster, and one sustained serious injuries (he did manage to indicate that he could not move). The first explosion damaged the mine's ventilation system, a system that was already deficient as evidenced by the 92 violations of ventilation standards issued by the Mining Safety and Health Administration during the 9 months prior to the disaster. With a disrupted airflow, the mine transitioned to a new hazardous state of methane accumulation (see Fig. 7).

Second accident pathogen: The second accident pathogen involves coal dust. Some background information is in order. Combustible dust is an insidious hazard with catastrophic potential across a broad spectrum of industries. Industries at risk of dust fires and explosions include food production, metal processing, wood, chemical manufacturing, and mining. Dust hazards are probably not well understood or their catastrophic potential not properly appreciated by some industry professionals, as evidenced by the repeated dust explosions in the United States – for example the Chemical Safety Board identified 281 combustible dust fires and explosions in the US between 1980 and 2005 and described this pattern of accidents as a "significant industrial safety hazard" (Combustible Dust, 2009). Consider for instance sugar dust. It may be surprising to think of sugar as an explosive hazard, but on 7 February 2008, a sugar dust explosion occurred at the Imperial Sugar Refinery near Savannah Georgia killing 14 workers, injuring 38 others, and almost completely destroying the plant. Whether Aluminum dust, saw dust, or coal dust for example, when confined in space and dispersed in the air in high concentration levels (within an explosive range or above a "Minimum Explosible Concentration"), they can become extremely potent accident pathogens waiting for an ignition source to be unleashed. In addition, dust explosions often occur in sequence, with a primary explosion unsettling dust accumulated in various places and dispersing it in the air, thus creating a new mixture and additional fuel load for secondary explosions. This particular accident mechanism allows dust explosions to propagate significantly further from the initial accident location, thus threatening a much larger area and more people than those within the vicinity of the primary explosion (the primary and secondary dust explosions can occur almost simultaneously or continuously).

Coal dust is a recognized explosive hazard in the mining industry. Explosions in coal mines involving coal dust are among the most violent and destructive type of accidents. The primary way of neutralizing this hazard, in addition to cleaning up and avoiding accumulation of coal dust in the mine and especially on electric equipment, is through rock dusting, that is the application of rock dust, a neutral dust to increase the amount of incombustible content in overall mine dust.[8] The US Mining Safety and Health Administration (MSHA) notes the following on the effectiveness of rock dusting:

"The law requires that all areas of a coal mine that can be safely traveled must be kept adequately rock dusted to within 40 feet of all working faces. *These are minimum requirements*. The chance of propagation and risk and widespread explosion disasters in coal mines can be nearly eliminated when rock dust is applied liberally and maintained properly" (Rock Dusting, 2010).

Going back to JWR disaster, rock dusting at the No. 5 mine was inadequate, and the hazard of dust explosion was probably not well appreciated at the mine, as evidenced by the 99 violations of coal dust accumulation and inadequate rock dusting issued in the 12-month prior to the accident (McKinney et al., 2002). In addition, during the 3 weeks prior to the accident, MSHA cited the mine operator 10 times for coal dust related violations, over an area covering 22,000 feet in the mine.

Thus, the second accident pathogen at the JWR No. 5 mine consisted in the accumulation of coal dust and the longstanding inadequacy of rock dusting at the mine. Coal dust became a major fuel source for the second explosion, and both the severity and the extent of propagation of the second explosion were the result of coal dust involvement.

Two fatal operational decisions: At this point in the accidence sequence, the first explosion had occurred (5:20 pm), mine ventilation was damaged and airflow was disrupted, methane was accumulating – recall the mine was characterized as and known to be "very gassy" – and coal dust was lurking over an extended part of the mine (see Fig. 7). Given these conditions, two fatal operational decisions precipitated the accident and aggravated its consequences. First, the block lights in the section were not de-energized [9]; they in effect provided the ignition source for the second explosion. Second, the mine was not ordered to be evacuated. The mine turned into a ticking bomb after the first explosion, a hazard properly recognized by some, and one person asked the control room supervisor if "all the miners were on their way out". However, instead of an immediate evacuation (immediately after 5:20 pm), miners remained underground and several were sent to the location of the first explosion. The decision not to order an immediate evacuation of the mine reflects poor training and deficient procedures for dealing with emergency situations. The accident report noted for example that "all the miners were not participating in fire drills every 90 days" as required by law and that responsibilities for dealing with mine emergencies were not clearly delineated (McKinney et al., 2002). In short, managerial and procedural deficiencies were present at the mine, and while they may have remained unobservable during normal operations, they surfaced at one crucial moment when the need for proper organizational decision-making was most acute following the first explosion. It is worth pointing out that while the second explosion may or may not have been prevented given the conditions at the mine, its consequences could have been significantly mitigated with proper training and procedures for organizational response to emergency situations. *Final state and consequences:* At 6:15 pm, 55 min after the first explosion, the "grace period" in this accident sequence during which decisions could have been made to mitigate the consequences of the accident, came to an end: a second more violent explosion involving methane and coal dust engulfed several parts of the mine and killed 13 miners.

The discussion in this section described the accident trajectory or sequence of the JWR No. 5 mine disaster. Several concepts were introduced to help describe the phenomenology of accidents in general and this accident in particular, such as *initiating event, precursor* or *lead indicator,* and *accident pathogen.* One common theme throughout this discussion is that factors contributing to an accident are not to be found only in the temporal vicinity of the moment when an adverse event occurs, but they can extend much further in the past. Factors that are controlled by human decisions and that can influence safety (not necessarily in an immediate or deterministic sense) are referred to as safety levers. These factors can be of different nature (technical, managerial/organizational, and regulatory) and they may contribute to accident prevention (prevention of initiating event), accident containment (blocking an accident sequence at the intermediary events), or consequence mitigation (the accident would occur but its consequences are minimized). In the following section, we expand on these ideas and introduce the general safety strategy of defense-in-depth. We then revisit JWR No. 5 mine disaster and examine the relevance of defense-in-depth for this particular accident and more broadly for thinking about safety in the mining industry.

DEFENSE-IN-DEPTH AND SAFETY LEVERS

Defense-in-depth is a fundamental principle/strategy for achieving system safety. First conceptualized within the nuclear industry, defense-in-depth is the basis for risk-informed decisions by the US

Nuclear Regulatory Commission (Sorensen et al., 1999), and is recognized under various names in other industries (e.g., *layers of protection* in the chemical industry (Layers of Protection Analysis, 2001, Kletz, 1999 and Summers, 2003). Accidents typically result from the absence or breach of defenses or violation of safety constraints (Leveson, 2004, Rasmussen, 1997 and Svedung and Rasmussen, 2002). Defense-in-depth embodies the idea of multiple lines of defense or safety barriers along accident scenarios, and it shuns the reliance of safety on a single element (hence the "depth" qualifier).

Defense-in-depth, typically realized by successive and diverse safety barriers, technical, organizational, and procedural, is designed to: (1) prevent incidents or accident initiating events from occurring, (2) prevent these incidents or accidents initiators from escalating should the first barriers fail, and (3) mitigate or contain the consequences of accidents should they occur (because of the breach or absence of the previous "prevention" barriers) (Sorensen et al., 1999).

It is worth pointing out that an accident is defined by the Department of Energy (Implementation Guide for Use with DOE Order 225.1A, 2010) as an "unwanted transfer [or release] of energy that, due to the absence or failure of barriers and controls, produces injury to persons, damage to property, or reduction in process output." This definition reflects the fundamental energy model of accidents, and the safety strategy of defense-in-depth is intrinsically related to this "energy model" of accidents. In essence, defense-in-depth is meant to prevent, mitigate, or contain unwanted releases of energy.

In mining as well as in other hazardous industries, various safety levers exist and can be acted upon to modify the exposure to the inherent risks involved in said industries, and promote safety. Fig. 8 provides an illustrative representation of various types of safety levers, along with examples of different stakeholders in the safety value chain.[10] Details on these concepts can be found in Saleh et al. (submitted for publication).

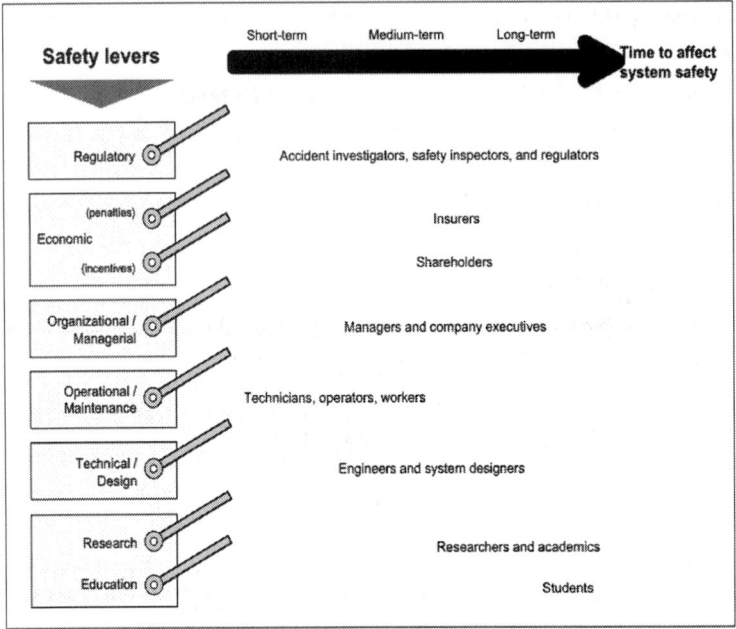

Figure 8: Safety levers and stakeholders in the safety value chain.

The importance of the safety principle of defense-in-depth cannot be underestimated. How the safety levers are pulled and what "defenses" have been put in place along potential accident trajectories – that is ways in which hazards transform into accidents – are essential contributions for the understanding of accident causation and the support of system safety.

For example, the regulatory lever, which historically has been an effective safety lever in the US mining industry, as discussed in Section 2, can be pulled in a variety of ways, for instance:

- By mandating pertinent (and up-to-date) technical standards and safety procedures, supported by proper research and epidemiological studies.

- By providing the inspection apparatus of the regulatory agency with the necessary resources – in terms of funding and skilled inspectors – to audit's the industry's compliance with the regulation. In the case of the mining industry, the passage of safety standards, for example with respect to mine ventilation, methane concentration monitoring, rock dusting, and safety training, is an important first step in pulling the regulatory safety lever. However, the effectiveness of such action is largely contingent on the ability to inspect the implementations of these safety standards in mines.[11]

- By supporting the enforcement capability of the regulatory agency, with a proper legal framework, to enforce compliance with the safety standards, deter or redress safety violations, and if need be, prosecute repeated safety violations.

These three types of actions on the regulatory safety lever, Mandate – Inspect – Enforce, can be complemented in a variety of ways, by incentivizing compliance for example, or by assisting the industry in building its skills and ability to comply with the mandated safety standards. The regulatory safety lever can encompass a wide range of actions and predispositions: one the one hand, detection and deterrence through the ability to escalate to threats and legal action against criminal or reckless non-compliant behavior; and on the other hand, incentives and assistance to comply with safety standards, especially for companies that have the intent but not necessarily the ability and required (organizational and technical) competence to comply (Ayers and Braithwaite, 1992).

It is useful to acknowledge, although beyond – or on the edge of – the scope of the present work, that safety regulations and their enforcement, and more broadly the regulatory process, are significantly shaped, among other things, by the political context. For example in the US, Scholz and Wei (1986) found clear evidence of differences in regulatory enforcement activities of the Occupational Safety and Health Administration (OHSA) in response to political factors at both the Federal and state levels. One of the findings of this study for example was that states with Democratic governors or legislatures tended to have more frequent workplace safety violations, and larger penalties for these violations, than their Republican counterparts. This important study concluded that regulatory bureaucracy, and more specifically its enforcement activities, responds to political demands but does so in a complex way, by "integrating political demands at various levels [and by] adapting central policies to fit into varied and changing local conditions," economic and other.

Kagan (2004) provides and excellent synthesis of "protective regulations" and the regulatory process. He notes that:

Partisan electoral politics has been shown to affect regulatory policies and enforcement methods. As the cost imposed by the regulatory state has grown, conservative political parties often promise to reduce the regulatory burdens on the business sector, while left-of-center parties typically promise to make regulation more stringent and effective.

He then proceeds to illustrate how partisan politics can affect the regulatory safety lever shown in Fig. 8:

Once elected, political party leaders affect [regulatory] agencies' policies and enforcement methods in many ways – by appointing [...] top agency officials; by expanding or contracting agency staffing and resources through the budget process; by legislative oversight hearings; and sometimes by telling regulatory officials how they would like regulatory officials how they would like regulatory issues of urgent political concerns to be handled.

Other ways political factors can shape the regulatory process in the US is through executive reviews and the issuance of Presidential Executive Orders (EO).[12]Kagan (2004) discusses the dilemmas of regulatory enforcement[13] and the different styles of regulations in the US and abroad (legal/adversarial process versus social/cooperative process). He indicates that effective regulators have at their disposal credible legal sanctions and the possibility to escalate up a "pyramid of sanctions" to meet an enterprise's repeated noncooperation and safety violations, and that socio-legal scholars tend to agree that the preferred regulatory style is flexible:

legalistic and punitive when needed, but accommodative and helpful in others, depending on the reliability of the regulated enterprise, and the seriousness of risks or harms created by particular violations.

The purpose of this discussion is not prescriptive, but to acknowledge the existence of multiple safety levers, and that each can be pulled or relaxed (weakened) in a variety of ways.

Defense-in-depth, and the commitment to its implementation, forces the thinking about regulatory, design, and operational choices to address various hazards and potential accident scenarios. Traditional risk analysis, in its bare essence, addresses the following questions (Apostolakis, 2004):

- What can go wrong?
- How likely it is?
- What are the consequences?

Defense-in-depth adds the most important complement to these questions, namely

- What are you doing about it ("it" being the answer to 1)? Or how are you defending against it?

The answer to this question constitutes an explicit demonstration of how various hazards and identified accident scenarios are, or ought to be, handled. In the next section, we examine how defense-in-depth can be relevant to the mining industry.

DEFENSE-IN-DEPTH AND THE MINING INDUSTRY

Several hazards affect the mining industry and render it one of the most dangerous worldwide. These include explosions (methane, coal dust, or other), fires, rock and roof falls, landslides, blackdamp and toxic gases outbursts, and water inrush/sudden inundations. Example of some of mining accidents in which these hazards turned into disasters are provided in Appendix A.

These hazards however can be prevented from escalating into accidents – or if they do escalate, their consequences can be contained or mitigated – with appropriate defenses. In other words, between "hazards" and "accidents", opportunities exist to prevent the transition of the former to the latter, and if this transition does occur, opportunities exist to prevent the transition of "accident" to "disaster" (consequence mitigation). The opportunities to control hazards and block accident trajectories can be captured by the safety principle of defense-in-depth and its implementation.

In order to have an effective control of hazards and to establish a set of "defenses" for blocking accident trajectories, it is important to understand first the "ingredients" that support the transition of the hazards into accidents, second the dynamic nature of this transition or the speed of development of a hazard into an accident, and third the "signatures" that a situation is growing into an increasingly hazardous state – what was referred to previously as the precursors or lead indicators. Once this understanding is established for all identified hazards, technical defenses, organizational and managerial defenses, and regulatory defenses can be put in place in support of safety in the mining system.[14]

The role of the miner is central to this proposed defense-in-depth approach to mine safety. The previous paragraph noted that for proper hazard control, it is essential to understand the ingredients of hazard build-up and escalation, as well as the "signatures" of these hazardous states and transitions—the operational recognition and awareness that an accident sequence may be unfolding. What was not indicated previously is where or in whom this understanding should reside; it is essential that this understanding be shared by all miners, supervisors, and mine management (not just with safety inspectors for example or other individuals remotely connected to the mine). The reasons for this statement may be self-evident but they are worth making explicit herein. The miners are at the "sharp-end of safety" (Reason, 1997), and being the closest to the potential hazards, their role in defense-in-depth is essential and covers three broad types of contributions:

- Miners are the principal agents in defense-in-depth. The shared knowledge of the ingredients for hazard build-up and escalation invites their participation in, or contribution towards, eliminating these ingredients and the prevention of accident pathogen build-up. In addition, it is the miners' active intervention in some cases that can help de-escalate an unfolding accident sequence and bring a mine back from a hazardous state to nominal conditions, when possible.

- Miners are also crucial "sensors" in defense-in-depth: many hazardous conditions in the mines and "signatures" of an increasingly dangerous state are best monitored and identified by the miners. In other words, miners can fulfill the essential role of a (distributed) monitoring network of local hazardous conditions within a mine. In short, they monitor for hazardous states and provide the prerequisite information for the triggering of active safety interventions in lines of defenses.

For these two roles to be properly fulfilled, mine management has to organize the necessary safety workshops for all miners to participate in, and during which the ingredients of various mine hazards are discussed, the dynamics of hazards escalation are laid out, and the "signatures" of various hazardous states are identified. These dedicated safety workshops can be organized on a regular basis to maintain an active awareness of mining hazards and a tested competence in identifying and dealing with them. These workshops can also help bring the collective wisdom (and experience) of miners to bear on finer details of local conditions, and other hazardous issues that the workshop organizers/facilitators may be unaware of or have overlooked. Finally, these workshops can have a strong positive side effect, namely the signaling by management the serious commitment of the company to safety (beyond the common lip service "safety first"). Management should promote safety vigilance and help build and sustain safety competence in all employees.

- Finally, safety-competent miners fulfill a crucial role, which can be termed decentralized decision-making in support of accident mitigation or containment, a final line of defense, especially during emergencies when centralized decision-making is absent, unavailable, or flawed (as was the case in the No. 5 mine case study, and many other mine disasters). During nominal operating conditions in the mine, a strong

- The second flawed line of defense was the inadequate training in understanding roof fall hazards, in dealing with this hazard, and interpreting its precursor signs. For example, proper training and procedures could have been put in place for aggressive response to deteriorating roof conditions, such as proper information sharing about the condition across the mine system (including management and mine engineer(s), as well as cordoning the area and de-energizing the area-at-risk.

Proper training would have resulted in either the battery never being brought to the area-at-risk (of roof fall) in the first place or that it would have been covered or removed at the signs of impending roof fall (see discussion of lead indicators in Section 4).

Furthermore, several lines of defense failed and allowed a serious accident pathogen, coal dust, to accumulate in the mine, in particular the failure of procedures for and implementation of rock dusting. In addition, the fact that the block lights remained energized after the first explosion and that the mine was not evacuated reflect important deficiencies in management-driven safety defenses of proper training and supervision. Training for emergency situations is or should be taken as a significantly important safety defense.

Finally, the regulatory safety lever, which creates several lines of defense, failed to enforce the implementation of proper technical safety barriers at the mine, as evidenced by the staggering 90+ safety violations of ventilation standards, and over 90 violations of coal dust accumulation and inadequate rock dusting in the year prior to the accident. In short, the regulatory safety defenses were inefficient and failed to prompt management to take safety seriously and implement the required technical and procedural defenses in support of mine safety. Why this occurred and safety inspections came to be considered as the "toothless tiger" (Jim Walter Resources #5 Mine Disaster, 2010) is an important topic for MSHA leadership and mine inspectors to analyze and reflect upon. The potential effects of these lines of defense are illustrated figuratively in Fig. 9.

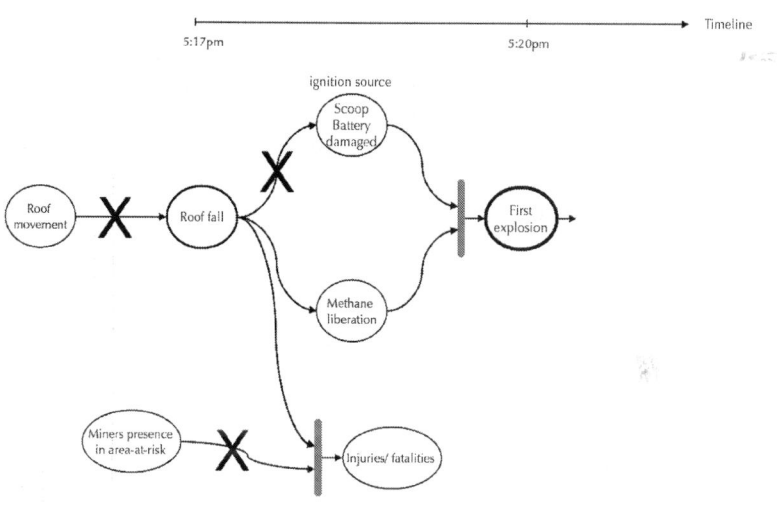

Figure 9: Illustrative effects of lines of defenses on a subset of the accident trajectory at the JWR No. 5 mine.

The process-centric relevance of defense-in-depth: Beyond the importance of the specific implementations of the principle of defense-in-depth for ensuring mine safety, we believe/hypothesize that this safety principle has several process-centric benefits. These include the following:

- Defense-in-depth will make it easy to organize safety trainings and get buy-in from miners and mine management for the various safety defenses that should be implemented.
- Safety workshops at mines will result in better information sharing and retention about particular hazards, their potential escalation, and the ways to defend against them, when the workshops are organized around the principle of defense-in-depth (for each particular hazard) than when the workshops consists of unstructured and increasingly larger lists of "DOs and DON'Ts".
- Defense-in-depth will support a better appreciation of accident lead indicators by miners and mine management and the need to strengthen lines of defense and trigger effective hazard control responses.

- Defense-in-depth will support a system approach to mining safety in which multiple stakeholders better understand their contributions to accident prevention through their actions on various safety levers and lines of defenses; mine safety becomes a collective output of coordinated efforts on various safety defenses.

MSHA, in collaboration with academia and select mining companies, should undertake a few pilot workshops at select mines to test these hypotheses and assess the appeal and usefulness (or lack thereof) of the establishment of defense-in-depth as the guiding safety principle for the mining industry.

CONCLUSIONS

In this work, we provided an overview of the broad and multi-faceted topic of safety in the mining industry. After reviewing some statistics of mining accidents, we focused on one pervasive and most deadly failure mode in mines, namely mine explosion. We analyzed one recent mine disaster in which various safety barriers failed to prevent the accident initiating event from occurring, then subsequent lines of defense failed to block this accident scenario from unfolding and to mitigate its consequences.

We then introduced the safety principle of defense-in-depth and examined how it can be relevant to and applicable in the mining industry in support of accident prevention and coordinating actions on various safety levers to improve mining safety.

Several important topics for the mining industry were intentionally left outside the scope of the present work (primarily because the length of the current article, and because a curt treatment would not do them justice). Some of these topics will be addressed in future work, and they include (1) mining health hazards – the present work focused solely on safety issues; (2) emergency response and post-accident analysis, information dissemination, and recommendations; (3) continuous risk analysis and identification of new hazards as conditions in the mining industry change; and (4) safety culture in the mining industry. Finally, we believe more interactions and partnerships between academia, beyond the traditional mining engineering discipline, and the mining community (mining operators, researchers, and regulators), would be particularly helpful in promoting safety innovations and advancing the safety agenda of the mining industry.

REFERENCES

1.   Apostolakis, G.E., 2004. How useful is quantitative risk analysis? Risk Analysis 24 (3), 515–520.

2.   Ayers, I., Braithwaite, J., 1992. Responsive Regulation: Transcending the Deregulation Debate. Oxford University Press, Oxford, UK.

3.   BBC News, 2007a. "Dozens dead in Ukraine mine blast". November 18, 2007. <http://news.bbc.co.uk/2/hi/europe/7100456.stm> (accessed 08.02.10).

4.   BBC News, 2007b. "Ukraine's mine death toll rises". November 20, 2007. <http:// news.bbc.co.uk/2/hi/europe/7103086.stm> (accessed 08.02.10).

5.   BBC News, 2010. "Scores rescued from flooded Chinese mine." April 5, 2010. http:// news.bbc.co.uk/2/hi/asia-pacific/8603102.stm (accessed 08.02.10)

6.   Black lung charts, 2009. Mine Safety and Health Administration, 2009. <http:// www.msha.gov/S./BlackLung/2009Charts/BlackLungCharts2009.pdf> (accessed 01.03.10).

7.   Blank, V.L.G., Andersson, R., Linden, A., Nilsson, B.-C., 1995. Hidden accident rates and patterns in the Swedish mining industry due to involvement of contractor workers. Safety Science 21 (1), 23–35.

8.   Brigg, E.F., 1964. Mine disaster. Science 146 (3640), 14. Combustible Dust, 2009. An Insidious Hazard. US Chemical Safety Board video. <http://www.csb.gov/videoroom/detail.aspx?VID=30> (accessed 01.03.10).

9.   CRS, 2007. Congressional Research Service. Changes to the OMB regulatory review process by Executive Order 13422. February 2007. Order Code RL33862.

10. Farmington Mine Disaster, 2010. United States Mine Rescue Association Homepage. <http://www.usmra.com/saxsewell/farmington.htm> (accessed 02.03.10).

11. Federal Mine Safety and Health Review Commission, 2005. Civil Penalty Proceeding, Secretary of Labor, Mine Safety and Health Administration vs. Jim Walter Resources, Inc., November 1, 2005. Docket No. SE 2003-160; AC No. 01–01322– 00004. Washington, DC.

12. GAO, 2009. United States Government Accountability Office. Federal rulemaking: improvement needed to monitoring and evaluation of rules development as well as transparency of OMB regulatory reviews. GAO-09-205.

13. Gates, R.A., Phillips, R.L., Urosek, J.E., et al., 2007. Report of investigation: fatal underground coal mine explosion, January 2, 2006. Sago Mine. Mine Safety and

14. Health Administration. http://www.msha.gov/sagomine/sagomine.asp (accessed 25.07.09).

15. Gates, R.A., Gauna, M., Morley, T.A., et al., 2007. Report of investigation: Coal Burst Accidents August 6 and 16, 2007. Crandall Canyon Mine. Mine Safety and

16. Health Administration. <http://www.msha.gov/Fatals/2007/Crandall Canyon/ Crandall Canyonreport.asp> (accessed 25.07.09).

17. History of Mine Safety and Health Legislation, undated. Mine Safety and Health Administration. http://www.msha.gov/MSHAINFO/MSHAINF2.HTM (accessed 03.15.11).

18. Hopkins, A., 1999. Managing Major Hazards: The Lessons of the Moura Mine Disaster. Allen & Unwin, Australia.

19. Hopkins, A., 2001. Was three mile island a 'normal accident'? Journal of Contingencies and Crisis Management 9 (2), 65–72.

20. Hugenard, P. (Ed.), 1996. Catastrophes: De la Strategie D'intervention a la Prise en Charge Medicale. Elsevier, Paris.

21. Implementation Guide for Use with DOE Order 225.1A, Accident Investigations, 2010. US Department of Energy. 1997, Washington, DC. <http:// www.directives.doe.gov/ directives/current-directives/225.1-EGuide-a-1/view> (accessed 15.02.10).

22. Jianjun, T., 2007. Coal mining safety: China's achilles' heel. China Security 3 (2), 36– 53.

23. Jim Walter Resources #5 Mine Disaster, 2010. United Mine Workers of America

24. Report. http://www.umwa.org/?q=content/jim-walters-resources -5-minedisaster (accessed 01.03.10).

25. Johnes, M., McLean, I., 2000. Aberfan Disaster. Nuffield College, University of Oxford, 2000 (based on the book "Aberfan: Government -@@- Disasters" by the same authors and published by Welsh Academic Press, Cardiff). <http:// www.nuffield.ox.ac.uk/politics/aberfan/home.htm> (accessed 26.02.10).

26. Kagan, R.A., 2004. Regulators and the regulatory processes. In: Sarat, Austin (Ed.), The Blackwell Companion to Law and Society. Malden, MA, USA, pp. 212–230.

27. Kelley, J.H., Kealy, D., Hylton Jr., C.D., Hallanan, E.V., Ashcroft, J., Murrin, J., et al., 1973. The Buffalo Creek Flood and Disaster: Official Report from the Governor's Ad Hoc Commission of Inquiry. West Virginia Division of Culture and History. http://www.wvculture.org/history/disasters/ buffcreekgovreport.html (accessed 26.02.10).

28. Kissell, F.N., 2006. Handbook for Methane Control in Mining. National Institute for Occupational Safety and Health. Pittsburgh Research Laboratory, Pittsburg.

29. Kletz, T.A., 1999. Hazop and Hazan: Identifying and Assessing Process Industry Hazards, fourth ed. Taylor & Francis, Philadelphia.

30. La Porte, T.R., 1996. High reliability organizations: unlikely, demanding and at risk. Journal of Contingencies and Crisis Management 4 (2), 60–71.

31. LaPorte, T.R., Consolini, P.M., 1991. Working in practice but not in theory: theoretical challenges of High-Reliability Organizations. Journal of Public Administration Research and Theory 1 (1), 19–48.

32. Launhardt, B., 2010. Sunshine Mine Fire: A View From The Inside. United States Mine Rescue Association Homepage. <http://www.usmra.com/saxsewell/   sunshine_view.htm> (accessed 02.03.10).

33. Layers of Protection Analysis, 2001. Simplified Process Risk Assessment. American Institute of Chemical Engineers (AIChE): Center for Chemical Process Safety, New York.

34. Leveson, N., 2004. A new accident model for engineering safer systems. Safety Science 42 (4), 237–270.

35. Martin, C.D., Maybee, W.G., 2000. The strength of hard-rock pillars. International Journal of Rock Mechanics and Mining Sciences 37 (8), 1239–1246.

36. McKinney, R., Crocco, W., Stricklin, K.G., Murray, K.A., Blankenship, S.T., Davidson, R.D., et al., 2002. Report of Investigation Fatal Underground Coal Mine Explosions September 23, 2001, No. 5 Mine Jim Walter Resources, Inc. Mine Safety and Health Administration. <http://www. msha.gov/fatals/2001/jwr5/ ftl01c2032light.pdf> (accessed 01.03.10).

37. Mining Disasters – An Exhibition, 1972. Sunshine Mining Company Mining disaster. Mine Safety and Health Administration.   <http://www.msha.gov/DISASTER/ SUNSHINE/SS2.asp> (accessed 01.03.10).

38. National Research Council, 2007. Mining Safety and Health Research at NIOSH, 2007. Reviews of Research Programs of the National Institute for Occupational Safety and Health. The National Academies Press, Washington, DC.

39. Perrow, C., 1984. Normal Accidents: Living with High-Risk Technologies. Princeton University Press, New Jersey.

40. Ramani, R.V., 1995. Mining disasters caused and controlled by mankind: the case for coal mining and other minerals Part

1: causes of mining disasters. Natural Resources Forum 19 (3), 233–242.

41. Rasmussen, J., 1997. Risk management in a dynamic society: a modelling problem. Safety Science 27 (2–3), 183–213.

42. Reason, J.T., 1997. Managing the Risks of Organizational Accidents. Ashgate, Aldershot, Hants, England; Brookfield, VT, USA.

43. Roberts, K.H., 1990a. Managing High-Reliability Organizations. California Management Review 32 (4), 101–113.

44. Roberts, K.H., 1990b. Some characteristics of one type of high reliability organization. Organization Science 1 (2), 160–176.

45. Rock Dusting, 2010. Mine safety and health administration. <http://www.msha.gov/   s&hinfo/rockdusting/rockdusting.asp> (accessed 01.03.10).

46. Saleh, J.H., Marais, K.B., Bakolas, E., Cowlagi, R.V., 2010. Highlights from the literature on system safety and accident causation: review of major ideas, recent contributions, and challenges. Reliability Engineering and System Safety 95 (11), 1105–1116.

47. Scholz, J.T., Wei, F.H., 1986. Regulatory enforcement in a federalist system. American Political Science Review 80 (4), 1249–1270.

48. Sorensen, J.N., Apostolakis, G.E., Kress, T.S., Powers, D.A., 1999. On the Role of Defense in Depth in Risk-Informed Regulation. International Topical Meeting on Probabilistic Safety Assessment, Washington, DC, August 22–26, 1999.

49. Strengthening coal mine safety standards in China, 2007. United Nation Development Programme. http://www.undp.org.cn/projects/53962.pdf (accessed 10.03.10).

50. Summers, A.E., 2003. Introduction to layers of protection analysis. Journal of Hazardous Materials 104 (1–3), 163–168.

51. Svedung, I., Rasmussen, J., 2002. Graphic representation of accident scenarios: mapping system structure and the causation of accidents. Safety Science 40 (5), 397–417.

52. The Work-Related Lung Disease Surveillance Report, 2007. National Institute for Occupational Safety and Health Publication No. 2008-143. <http:// www.cdc.gov/niosh/docs/2008-143/default.html> (accessed 10.03.10).

53. Weick, K.E., Roberts, K.H., 1993. Collective mind in organizations – heedful interrelating on flight decks. Administrative Science Quarterly 38 (3), 357–381.

54. Weick, K.E., Sutcliffe, K.M., 2001. Managing the Unexpected: Assuring High Performance in An Age of Complexity. Jossey-Bass, San Francisco.

Fire and Explosion Hazards Related to the Industrial Use of Potassium and Sodium Methoxides

Q. Kwok[a], B. Acheson[a], R. Turcotte[a], A. Janès[b], and G. Marlair[b]

[a]Natural Resources Canada, Canadian Explosives Research Laboratory, 1 Haanel Drive, Bells Corners Complex – Building 12, Ottawa, Ontario, Canada K1A 1M1

[b]Institut National de l'Environnement Industriel et des Risques, Parc Technologique ALATA, BP 2, F60550 Verneuil-en-Halatte, France

ABSTRACT

Sodium and potassium methoxides are used as an intermediary for a variety of products in several industrial applications. For example, current production of so called "1G-biodiesel" relies on processing a catalytic reaction called "transesterification". This reaction transforms lipid resources from biomass materials into fatty acid methyl and ethyl esters. 1-G biodiesel processes imply the use of methanol, caustic potash (KOH), and caustic soda (NaOH) for which the hazards are well characterized. The more recent introduction of the direct catalysts CH_3OK and CH_3ONa may potentially introduce new process hazards. From an examination of existing MSDSs concerning these products, it appears that no consensus currently exists on their intrinsic hazardous properties. Recently, l'Institut National de l'Environnement Industriel et des Risques (France) and the Canadian Explosives Research Laboratory (Canada) have embarked upon a joint effort to better characterize the thermal hazards associated with these catalysts. This work employs the more conventional tests for water reactivity as an ignition source, fire and dust explosion hazards, using isothermal nano-calorimetry, isothermal basket tests, the Fire Propagation Apparatus and a standard 20 L sphere, respectively. It was found that these chemicals can become self-reactive close to room temperature under specific conditions and can generate explosible dusts.

INTRODUCTION

The commercial importance of sodium methoxide (CH_3ONa) and potassium methoxide (CH_3OK) has increased with the booming development of biodiesel production [1]. These chemicals are of major interest as catalysts, when compared to the "1G" catalysts such as caustic potash (KOH) or caustic soda (NaOH) for the so-called transesterification of triglycerides into fatty acid methyl or ethyl esters.

Like other alkyl metal oxides, these chemicals should be considered hazardous due to their reactivity with air, water, and carbon dioxide which may induce fire and explosion hazards. These assumptions can be made a priori from the information contained in MSDSs for both chemicals, as well as from existing hazards qualifiers from conventional classification systems and biodiesel related safety reviews [2].

As shown in Table 1, a review of the existing MSDSs for sodium methoxide [3], [4], [5], [6], [7] and [8] reveals that no consensus exists on several of its intrinsic hazardous properties.

Table 1: Available MSDS information on the hazardous properties of sodium methoxide

Auto-ignition temperature	From 50 °C [3] to 454 °C [4] and [5]
Decomposition temperature	From 50 °C [6] to 126 °C [3] and [7]
Reactivity with water	From incompatible [7] to violent [3], [4], [5] and [6]
Reactivity with air	Reacts with moist air [4] and [7], ignites in moist air [6] and [8]
Flammability (fire hazard)	From not flammable [5] to highly flammable [3], [6] and [8] to extremely flammable [7]
Explosive atmosphere	In combination with air [3], upon decomposition [7], in combination with metals [6] and [7]
ESD sensitivity	From "yes" [3] to "dusts only" [4] and [8]

The situation is quite similar for potassium methoxide, although much less MSDS information could be easily found in this case [9], [10] and [11]. Similar qualifiers are being used in conventional classification systems such as the European Regulation called "CLP" [12], which operates as the legal European implementation tool of the Globally Harmonized System (UN GHS) or the United Nations Recommendations on the Transport of Dangerous Goods (UN-TDG) [13]. Under the UN-TDG scheme, these chemicals are classified under class/division 4.2 as substances liable to spontaneous combustion. From views expressed to some of the authors by their potential distributors and users, it is known that some incidents involving these chemicals have occurred in the

procedure recommended by the manufacturer [23] was verified using a standard reaction [24] and [25].

Experiments with approximately 200 mg of sample, were initially attempted in a stainless steel sample vessel opened at the top to allow for water additions to be made after the sample was equilibrated into the calorimeter. However, in these experiments, it was impossible to achieve a stable baseline. Invariably, the calorimeter detected heat flow from the sample without any water being added. This was thought to be due to either the availability of room air in the sample vessel (reactivity with moist air [22]), or to a possible reaction with the metallic sample vessel itself.

Another series of experiments was therefore performed on dry sodium methoxide by introducing the sample (100 mg) in a glass sample vial prepared in a glove box under argon atmosphere. The sample was then sealed in the vial, still under argon atmosphere, using a crimped cap having a rubber septum. The sample vial was then introduced into the calorimeter and left to equilibrate for several hours. Using this procedure, a steady baseline was achieved. Following equilibration of the baseline, successive water additions were performed using a 100 μL syringe equipped with a long needle pushed through the rubber septum of the sample vial.

IBT

Critical self-ignition temperatures in different sizes of cubic baskets were determined by successive tests in a heating oven maintained at a fixed temperature, according to the standard EN 15188 [26]. From the latter, critical self-ignition temperatures are assessed using 125 L re-circulating air oven, with a maximum temperature of 300 °C, and equipped with a middle-height grating, as shown in Fig. 1. Ventilation was provided to exhaust any gases released from the powder sample.

Figure 1: View of a filled basket inside the IBT heating oven.

The powder sample was filled into cubic mesh wire baskets of different sizes with a spatula without compressing, and then levelled at the top. The baskets are open at the top and closed at the bottom. The width of the stainless steel mesh is equal to 10 μm in order to avoid sifting but to allow ambient air diffusion through the sample. Basket volumes of 8 and 125 cm³ were used in the present work. These baskets, filled with sample material, were positioned at the centre of the oven already preheated to the test temperature (Fig. 1). During the tests, the oven temperature was monitored using a thermocouple freely installed in the air space at half distance between the sample surface and the inner wall of the oven. Tests have shown that the maximum temperature difference between two points in the oven is of the order of 2 °C. The sample temperature was also monitored using a thermocouple with its hot junction imbedded in the centre of the sample.

In accordance with the above EN standard, if the sample temperature, after some induction delay time, increases abruptly and rises at least 60 °C above the oven temperature, it is considered that self-ignition has occurred. On the contrary, if the sample temperature remains close to that of the oven or exceeds it by only a few degrees after some induction time, it is assumed that self-ignition did not occur within this time period.

FPA

The INERIS' FPA instrument was used to determine the fire behaviour of sodium and potassium methoxides and their solutions in methanol. This apparatus (see schematic view in Fig. 2), based on standard ASTM E 2058 [27], was implemented in INERIS laboratories with the collaboration of FM Global in 1997. Whereas originally the FPA was mainly devoted to study the fire behaviour of building materials, in recent years INERIS has developed significant experience making use of this equipment to determine the fire safety profiles of various chemicals.

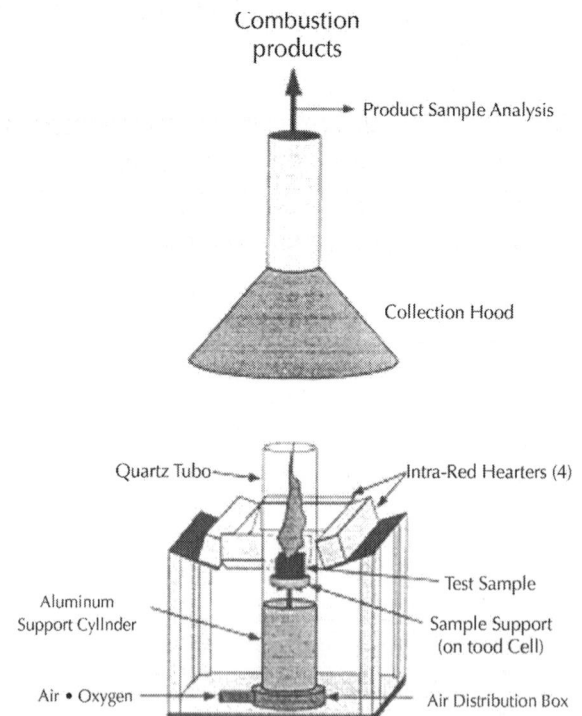

Figure 2: Schematic view of the Fire Propagation Apparatus.

The FPA operated by INERIS is slightly modified as compared with the standard configuration, in order to enhance safety of the operators during experiments on toxic chemicals. As with the standard configuration, parameters for standard "fire" under both ventilated and under-ventilated conditions may be set. With usual configurations, dry air is supplied in the bottom part of the system for perfect control of fuel/air ratio. This FPA comprises two main sections: a combustion area, and an exhaust (combustion gases, soot, etc.) line for collection and measurement of products. The test sample is placed on a holder mounted on a load cell and enclosed in an infrared transparent quartz tube delimiting the combustion area. With this set-up, the sample can also be subjected to a pre-determined external heat flux using four infrared heaters fixed at the exterior of the quartz tube. The heating system is both air and water cooled due to the high temperatures developed by the lamps. The main adjustment consists of a supply of dilution (or quenching) secondary air in a closed and gas-tight configuration between the two main parts of the calorimeter, making use of an annular ring shaped distributor to supply external air. More detailed descriptions of the apparatus can be found elsewhere [28] and [29].

As compared with the standard configuration, the INERIS equipment is also considerably instrumented with gas measurement devices including; CO/CO_2, THC (total hydrocarbons), NOx, SO_2, O_2 on-line analysers, and an FTIR instrument. The details of conventional on-line analyser instruments (e.g. Non-Dispersive Infra-Red (NDIR) analyser for CO/CO_2 species) used with the FPA can be found in an earlier paper describing the INERIS FPA [30]. The FTIR instrument is a Fischer Rosemount model Nicolet 6700 equipped with a Mercury–Cadmium–Tellurium (MCT) type detector cooled with liquid nitrogen. Further details regarding the configuration of the instrument are also provided elsewhere [31].

In the measurements with the FPA instrument, approximately 50 g of each sample were laid in a sample holder (88 mm diameter quartz pan) and submitted to pilot ignition, under a 25 kW external heat flux. Mass loss, combustion gas emissions, and O_2 consumption for heat release measurements were also recorded.

Dust Explosivity

The dust explosivity of the "as-received" methoxide powders was measured in a standard 20 L sphere, in accordance with standard EN 14034 [32]. In this test, the powder to be dispersed in the main test vessel was stored in a 1 L reservoir. It was dispersed into the sphere, under partial vacuum (−400 mbar relative), through a perforated disperser, in order to obtain a powder cloud at atmospheric pressure. Piloted ignition was induced, 60 ms after the powder dispersion, by two chemical igniters installed at the centre of the sphere prior to the test. Each igniter was capable of delivering 5 kJ of energy.

During these tests, the pressure-time history of the event was recorded and, from this record, the maximum overpressure (Pm) and the maximum explosion pressurization rate (dP/dt)m were computed. The tests were repeated by varying the mass of powder and therefore the average powder concentration in the test vessel. Then, the particular value of the concentration for which Pm and (dP/dt)m reached their maximum values P_{max} and $(dP/dt)_{max}$ was determined from the average values bracketing the optimal concentration. From these experimental values, the explosion index (K_{st}) [32] can be calculated, according to the following relation:

$$K_{st} = \left(\frac{dP}{dt}\right)_{max} V^{1/3} \, (\text{bar m s}^{-1})$$

where V is the volume of the sphere (m^3). Determination of the K_{st} parameter allows one to assign a hazard class to the tested dust.

RESULTS

INC

A typical measurement obtained for sodium methoxide using a sealed glass sample vessel under argon atmosphere is shown in Fig. 3. In this case, water was added 10 µL at a time. Each water addition is clearly followed by a fast heat flow signal with enough waiting time in between additions, for the heat flow to return approximately to the original baseline. Integration of the exothermic peaks from the INC experiment of Fig. 3, results in an approximate heat of reaction of -160 J g^{-1} for sodium methoxide with water. This value was obtained by summing up the integration of each individual peak, after baseline correction. Due to the fact that for the first three water additions, the detector response was slightly saturated, the actual heat evolved may have been larger than that the instrument could measure. The above value is therefore an approximate measure of the heat of reaction of sodium methoxide with water; a reaction which produces primarily methanol and sodium hydroxide. After the addition of 60 mass % of water, however, the response of the sample became endothermic (see Fig. 3). It was verified that the addition of water to an empty INC sample vial resulted in an endothermic response of the calorimeter as well. Therefore, this endotherm was taken as an indication that the exothermic reaction had proceeded to completion.

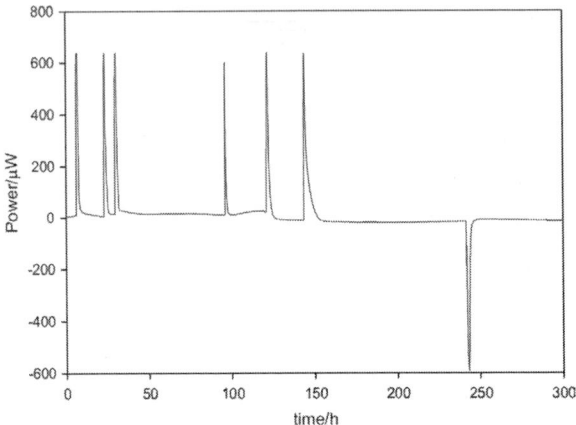

Figure 3: INC heat flow curve for seven 10 μL water additions to 102 mg of dry CH₃ONa under argon atmosphere.

IBT

The results from the isothermal basket tests for both methoxides in 8 cm³ baskets are shown in Fig. 4, while those obtained for sodium methoxide in a 125 cm³ basket are displayed in Fig. 5.

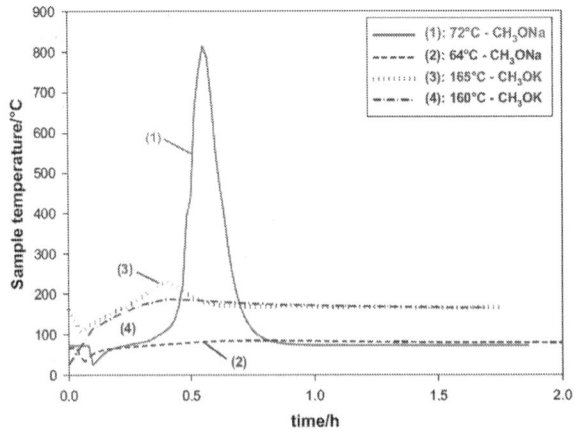

Figure 4: Results of the 8 cm³ isothermal basket tests for both CH₃ONa and CH3OK.

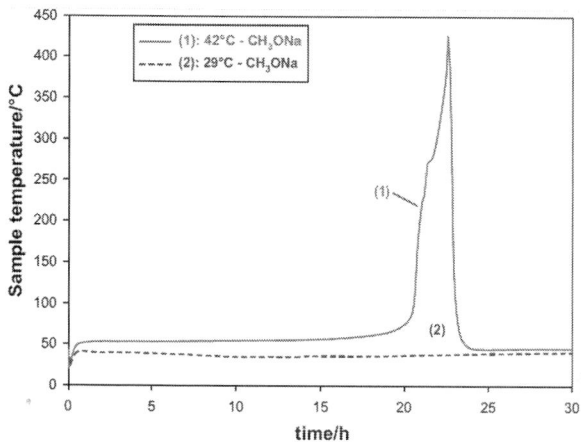

Figure 5: Results of the 125 cm³ isothermal basket tests for CH₃ONa.

With the 8 cm³ basket, sodium methoxide showed no self-heating at an isothermal temperature of 64 °C for up to 2 h, but a clear self-ignition was observed at an isothermal temperature of 72 °C in less than 0.5 h. According to this result and the author›s experience with basket testing, examination of self-heating process at 64 °C was stopped after 2 h for 8 cm³ baskets. Therefore, for this sample size, a critical temperature of self-ignition of about 65–70 °C was obtained for sodium methoxide. For potassium methoxide, no self-heating was observed at 160 °C, while some self-heating was detected at the isothermal temperature of 165 °C. This somewhat agrees with the results of large scale thermogravimetry-differential thermal analysis (TG-DTA) experiments performed on the same samples [22].

With the 125 cm³ basket, no self-ignition was detected for sodium methoxide at an isothermal temperature of 29 °C for up to 5 days. Self-ignition occurred at an isothermal temperature of 42 °C after an induction time of about 20 h. On the basis on these results, a critical temperature for self-ignition between 29 and 42 °C was obtained for the 125 cm³ sample of sodium methoxide. This clearly demonstrates a potential self-heating hazard during industrial storage under normally expected ambient storage conditions when in contact with air.

FPA

With methoxide solutions in methanol, the determined CO_2 yields were found to be comparable to those of neat methanol (about 1400 mg g^{-1}), which is very close to the theoretical maximum yield for methanol (1375 mg g^{-1}). However, the CO_2 emission curves revealed an unexpected behaviour deviating from that of a pure methanol solution. As illustrated in Fig. 6, a regular and gradual decrease in CO_2 release rate was observed for the CH_3ONa solution, compared to the typical plateau followed by a sharp decrease, which is normally observed in conventional liquid fires involving hydrocarbon molecules [33]. This difference in behaviour may be explained by the observation of solid precipitate forming on the surface of the pool at the early stage of the burning process. As seen in Fig. 7, this solid deposit tended to impede air diffusion to the solvent phase thus progressively reducing the combustion rate. By comparison, in the case of the CH_3OK solution, this surface precipitate was observed to appear at a later stage and much more suddenly, which was reflected by an abrupt change in CO_2 release rate profile, followed by an apparent extinction of the burning mechanism.

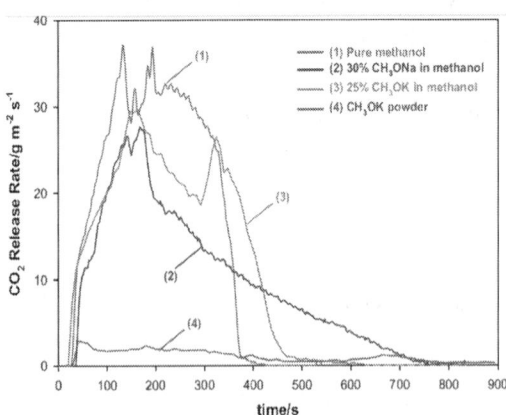

Figure 6: CO_2 release rate of methoxide powder and methoxide solutions compared to that of methanol in the FPA apparatus.

Figure 7: Precipitate forming process on the surface of a burning pool of sodium methoxide solution in the FPA.

The fire behaviour of the powdered methoxides was also consistent with these observations. In this case, very limited surface flaming was initially observed. This was followed by glowing combustion leading to hardly measurable mass loss and apparent extinction. Observation of the residues showed the formation of an air-tight solid interface. However, when the interface of the residue was manually cracked, red hot-spots appeared due to fresh contact with air. It is possible that these residues may be unreactive alkali carbonates and alkali methyl carbonates ($KOCO_2CH_3$, $NaOCO_2CH_3$), as it is well known that alkali methoxides readily react with CO_2 exothermically at room temperature to produce these products [34].

Dust Explosivity

The evolution of Pm and (dP/dt)m as a function of dust concentration for the 20 L sphere experiments on both potassium and sodium methoxide is shown in Fig. 8 and Fig. 9, respectively. The results of a particle size analysis for these powders have shown that the average grain size was of the order of 100 μm in both cases.

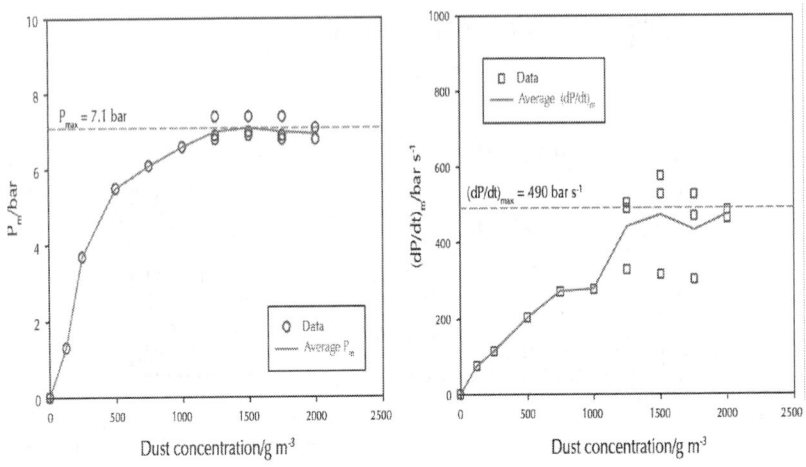

Figure 8: Evolution of Pm and (dP/dt) m as a function of dust concentration for CH_3OK.

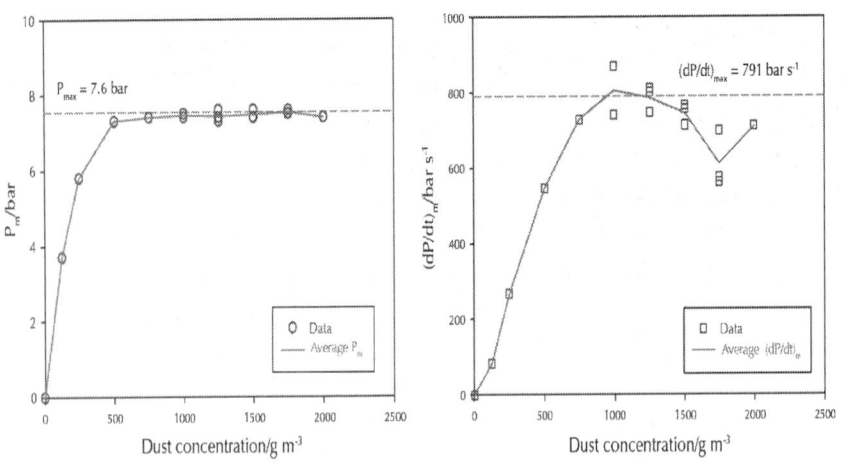

Figure 9: Evolution of Pm and (dP/dt) m as a function of dust concentration for CH_3ONa.

The calculated Kst values and corresponding hazard class assignments are provided in Table 2. These results can be used for the design of dust explosion protection equipment for the related industrial processes.

Table 2: Results of dust explosion characteristics for sodium and potassium methoxide

	P_{max}	$(dP/dt)_{max}$	K_{st}	Explosivity class
Potassium methoxide	7.1 bar	490 bar/s	133 bar m/s	St1 (1–200 bar m/s)
Sodium methoxide	7.6 bar	791 bar/s	215 bar m/s	St2 (201–300 bar m/s)

DISCUSSION

The results of the INC measurements show that only a limited amount of energy is released when sodium methoxide is put into contact with water (about 160 J g^{-1}) under an inert atmosphere. It is generally recognized that exothermic reactions releasing less than about 100–200 J g^{-1} rarely represent a runaway hazard [35]. However, this amount of energy could produce a sizable local temperature rise that might be capable of causing ignition of process fuels having low temperature flash points.

The results of the 8 cm^3 IBT tests on sodium methoxide were consistent with those of the previously reported large scale TG-DTA and ARC experiments for which the sample had access to sufficiently large quantities of moist air [22]. However, the results obtained for potassium methoxide show a much reduced response and much higher onset temperatures. This may be explained by the fact that for these tests, the samples were not stored under an inert atmosphere. They were kept in their respective "as-received" glass vials. Sodium methoxide was tested first, while potassium methoxide was tested several months later. It may well be that the potassium methoxide had been significantly aged, in which case a much reduced response would be expected, based on evidence obtained in DSC tests performed on samples aged for several hours [22].

The results of the 125 cm³ IBT tests on sodium methoxide clearly show that, as the sample mass is scaled up, the self-ignition (or onset) temperature approaches plausible storage temperatures when the sample is allowed to have intimate contact with ambient air. This demonstrates a very severe self-ignition (or runaway) risk for the associated industrial processes if the methoxides are allowed to come into contact with moist air.

As revealed from the FPA experiments making use of dry air, pool burning of methoxide solutions initially behaved like a typical methanol pool fire, until precipitation of white salts (possibly carbonates and methyl carbonates [34]) occurred. The solid white deposit appeared sooner in the case of the sodium methoxide solution as compared to potassium methoxide solution.

The appearance of this solid white deposit was observed to promote the formation of a crust, which caused an air-tight interface. However it was observed that hot-spots may survive for some time under this interface. These hot-spots may become ignition sources for flammable atmospheres induced by methoxide solutions in methanol or other fuels encountered in the associated manufacturing processes. Therefore, deposits of solid form of methoxides should be avoided throughout all process equipment, including ducting networks.

The results of the dust explosion tests show that, for the particular grain size distribution of the samples obtained from Sigma Aldrich, potassium methoxide dust is moderately explosible, while sodium methoxide dust is strongly explosible. Therefore, proper systems should be put in place to avoid ignition sources and/or provide adequate venting in the associated process equipment. It should be pointed out that these results are expected to be dependent on the physical properties of the powder samples. Therefore, methoxide batches originating from different suppliers may have different dust explosivity ratings.

CONCLUSIONS

The present work complements earlier work on the thermal properties of potassium and sodium methoxides [22] by providing data on the fire, explosion, as well as water reactivity hazards.

Whereas several authors [2], [3], [4], [5] and [6] have insisted on the high reactivity of these methoxides with water, the present results seem to demonstrate that oxygen and carbon dioxide play an important role in this reaction. Indeed, under inert atmosphere, it is found that less than 200 $J g^{-1}$ is released when water is put in direct contact with the methoxides. The results of the IBT tests are consistent with earlier thermal stability analysis [22]. The present results also clearly demonstrate that important self-heating is possible only slightly above room temperature.

Study of combustion behaviour in layers of methoxides has highlighted the formation of hot-spots that can survive for a long period of time. When these hot-spots are put into contact with fresh fuel, they can act as potential ignition sources. Finally, dust explosion tests show that methoxides dusts are explosible, with varying severities.

From some of the author's discussions with sodium and potassium methoxide users, it is known that the actual chemicals used in the biodiesel industry are less pure than the samples tested in the present work, and possibly physically different. This generally implies different grain size distribution and an increased presence of impurities. In the future, it would be imperative to better characterize these "industrial grade" methoxides so that the measurements can provide more representative estimates of the associated process hazards.

ACKNOWLEDGEMENTS

The authors would like to thank Ms. K. Basu, Ms. C. Johnson and Mr. J.P. Bertrand for their participation in the experiments reported herein. They would like also to thank the SAS PIVERT for its financial contribution in the frame of the French Institute of Excellence in the field of Low-Carbon Energies (IEED) P.I.V.E.R.T. (www.institut-pivert.com) selected as an Investment for the Future ("Investissements d'Avenir")

REFERENCES

1.   G. Marlair, P. Rotureau, H. Breulet, S. Brohez, Booming development of biodiesel for transport: is fire a safety concern? Fire Mater. 33 (2009) 1.

2.   Health and Safety Executive, Inspection of biodiesel sites ref. SPC/ENFORCEMENT/137, http://www.hse.gov.uk/ foi/internalops/hid/spc/ spcenf137.htm, 2012 (accessed 20.07.12).

3.   Green catalysts, sodium methoxide powder MSDS. http:// greencatalysts.com/SOMETH-msds.pdf, 2009 (accessed July 20.07.12).

4.   Anachemia, Material safety data sheet – sodium methoxide anhydrous. http://www.anachemia.com/msds/english/8416. pdf, 2010 (accessed July 20.07.12).

5.   Sigma–Aldrich, MSDS – sodium methoxide purum. http://www. sigmaaldrich.com/catalog/ProductDetail. do?D7=0&N5=SEARCH CONCAT PNO %7CBRAND KEY &N4=71750%7CFLUKA&N25=0&QS=ON&F=SPEC, 2010 (accessed July 20.07.12).

6.   IPCS Inchem, Sodiummethylate.http://www.inchem.org/ documents/icsc/icsc/ eics0771.htm, 2009 (accessed July 20.07.12).

7.   Avantor Performance Materials, Inc., MSDS – Sodium Methoxide. http://www.avantormaterials.com/documents/ MSDS/usa/English/M2028 msds us Default.pdf, 2011 (accessed 20.07.12).

8.   Science Lab.com, Inc., Material Safety Data Sheet – Sodium methoxide MSDS. http://www.sciencelab.com/msds. php?msdsId=9927332, 2012 (accessed 19.10.12).

9. GELEST Inc., MSDS Potassium Methoxide. http://www.gelest. com/GELEST/Forms/Search/ContentArea/ChemBioVizSearch. aspx?FormGroupId= 50000&AppName=GELEST&AllowFul lSearch=true&KeepRecordCount Synchronized=false&Que ryFormID=2&ListFormID=1&DetailFormID=0&Search Crite riaId=1&SearchCriteriaValue=potassium+methoxide, 2009 (accessed 19.10.12).

10. Sigma–Aldrich, MSDS – Potassium Methoxide, 2011, http:// www. sigmaaldrich.com/safety-center.html, 2011 (accessed 19.10.2012).

11. Alfa Aesar, MSDS – Potassium Methoxide. https://us.vwr.com/ stibo/hi res/AA14261-30 09232009.pdf, 2009 (accessed 19.10.2012).

12. Regulation (EC) n°1272/2008 of the European Parliament and the Council of 16 December 2008 on classification, labelling and packaging of substances and mixtures, amending and repealing Directives 67/548/EEC and 1999/45/EC and amending Regulation (EC) n°1907/2006.

13. UN Recommendations on the Transport of Dangerous Goods – Model Regulations, 15th revised edition, http://www. unece.org/trans/danger/ publi/unrec/GuidingPrinciples/ GuidingPrinciplesRev15 e.html, 2012 (accessed October 19.10.12).

14. S.W. Harper, J.C. Etchells, A.J. Summerfield, A. Cockton, Health & Safety in Biodiesel Manufacture, Hazards XX: Process Safety and Environmental Protection, Harnessing Knowledge, Challenging Complacency, in: Symposium Series 154, Institution of Chemical Engineers, December 2008, 2008, pp. 943–952.

15. Interstate Chemical Company, Inc., Sodium Methylate Handbook (contact information: Attn: Marketing Department, 2797 Freedland Road, Hermitage, PA 16148, Phone: 724-981-3771).

16. C. Rivière, G. Marlair, A. Vignes, Learning on the root factors of incidents potentially impacting the biofuel supply chains from

some 100 significant cases, in: G. Suter, E. De Rademaeker (Eds.), Proceedings of the 13th Int. Symposium on Loss Prevention and Safety Promotion in the Process Industries, June 6–9, 2010, Brugge, Belgium, 2010, pp. 35–42.

17. E. Salzano, M. Di Serio, and E. Santacesaria, Emerging Safety Issues for Biodiesel Production Plant, Paper presented at the 4th International Conference of Safety and Environment in the Process Industry, March 14–17, 2010, Florencia, Italy (see http://www.aidic.it/CISAP4/webpapers/106Salzano.pdf; accessed October 19, 2012).

18. S. Rivera, J.E. Nuniez Mc Leod, Human Error in Biofuel Plants Accidents, in: World congress on Engineering 2008, July 2–4, 2008, London, United Kingdom, 2008.

19. E. Salzano, M. Di Serio, E. Santacesaria, Emerging risks in the biodiesel production by transesterification of virgin and renewable oils, Energy Fuels 24 (2010) 6103.

20. S. Nair, Identifying risk and consequence assessment for major hazards and project design evaluation, in: HazardsXXII,April 11–14, 2010, Liverpool, United Kingdom., 2010.

21. http://igus-experts.org/, 2012 (accessed on October 15.10.12).

22. Q. Kwok, B. Acheson, R. Turcotte, A. Janès, G. Marlair, Thermal hazards related to the use of potassium and sodium methoxides in the biodiesel industry, J. Therm. Anal. Calorim. 111 (2013) 507.

23. User's Manual, Model CSC 4200, Calorimetry Sciences Corp., 2002.

24. C.M. Badeen, D.E.G. Jones, Performance, Assessment and Application of an Isothermal Nanocalorimeter, in: 5th Int. Symposium on the Heat Flow Calorimetry of Energetic Materials, September 12–15, 2005, Indianapolis, Indiana, USA, 2005.

25. A.E. Beezer, A.K. Hills, M.A.A. O'Neill, A.C. Morris, K.T.E. Kierstan, R.M. Deal, L.J. Waters, J. Hadgraft, J.C. Mitchell, J.A. Connor, J.E. Orchard, R.J. Willson, T.C. Hofelich, J. Beaudin, G. Wolf, F. Baitalow, S. Gaisford, R.A. Lane, G. Buckton,

M.A. Phipps, R.A. Winneke, E.A. Schmitt, L.D. Hansen, D. O'Sullivan, M.K. Parmar, The imidazole catalysed hydrolysis of triacetin: an inter- and intra-laboratory development of a test reaction for isothermal heat conduction microcalorimeters used for determination of both thermodynamic and kinetic parameters, Thermochim. Acta 380 (2001) 13–17.

26. European Standard NF EN 15188, Determination of the Spontaneous Ignition Behaviour of Dust Accumulations, European Committee for Standardization, Brussels, Belgium, 2007, ISSN 0335-3931.

27. AST.M.E. 2058-09, Standard Test Methods for Measurement of Synthetic Polymer Material Flammability Using a Fire Propagation Apparatus (FPA), American Society for Testing Material, West Conshohocken. PA, USA, 2009.

28. S. Brohez, G. Marlair, J.P. Bertrand, C. Delvosalle, The Effect of Oxygen Concentration on CO Yields in Fires, in: Proceedings of the 10th INTERFLAM International Conference, July 5–7, 2004, Edinburgh, Scotland. London: Interscience Communications, 2004, pp. 775–780.

29. H. Biteau, A. Fuentes, G. Marlair, S. Brohez, J.L. Torero, Ability of the fire propagation apparatus to characterize the heat release rate of energetic materials, J. Hazard. Mater. 166 (2009) 916–924.

30. G. Marlair, J.P. Bertrand, S. Brohez, Use of the Fire Propagation Apparatus for the evaluation of under-ventilated fires, in: Proceedings of the Fire & Materials Conference 2001, San Francisco, CA, 22–24 January 2001 Interscience Comm. Ltd., 2001, pp. 301–314.

31. D. Calogine, G. Marlair, J.P. Bertrand, S. Duplantier, J.-M. Lopez-Cuesta, R. Sonniers, C. Longuet, B. Minisini, C. Chivas-Joly, E. Guillaume, D. Parisse, Gaseous effluents from the combustion of nanocomposites in controlled-ventilated conditions, J. Phys. Conf. Ser. 304 (2011) 012019.

32. European Standard BS EN 14034-2, Determination of Explosion Characteristics of Dust Clouds. Determination of

the Maximum Rate of Explosion Pressure Rise (dp/dt)max of Dust Clouds, European Committee for Standardization, Brussels, Belgium, 2006, ISBN 0 580 48679 6.

33.   G. Marlair, C. Cwiklinski, F. Marlière, A review of large-scale fire testing focusing on the fire behaviour of chemicals, in: Proceedings of the 7th INTERFLAM International Conference, March 26–28, 1996, Cambridge (United Kingdom), Interscience Communications, 1996, pp. 371–382.

34.   B. O Heston, O.C. Dermer and J.A. Woodside, (1942), The Reaction of Alkoxide ions with Carbon Dioxide, Academy of Science for 1942, pp. 67-8.

35.   T. Grewer, Thermal Hazards of Chemical Reactions, Industrial Safety Series, vol. 4, Elsevier Science B.V, Amsterdam, 1994.

Influence of Hydroxyl Group Position and T emperature on Thermophysical Properties of Tetraalkylammonium Hydroxide Ionic Liquids with Alcohols

Pankaj Attri, Ku Youn. Baik, Pannuru Venkatesu,
In Tae Kim, and Eun Ha Choi

[1]Plasma Bioscience Research Center/Department of Electrical and Biological Physics, Kwangwoon University, Seoul, Korea

[2]Department of Chemistry, University of Delhi, Delhi, India, 3 Department of Chemistry, Kwangwoon University, Seoul, Korea

ABSTRACT

In this work, we have explored the thermophysical properties of tetraalkylammonium hydroxide ionic liquids (ILs) such as tetrapropylammonium hydroxide (TPAH) and tetrabutylammonium hydroxide (TBAH) with isomers of butanol (1-butanol, 2-butanol and 2-methyl-2-propanol) within the temperature range 293.15–313.15 K, with interval of 5 K and over the varied concentration range of ILs. The molecular interactions between ILs and butanol isomers are essential for understanding the function of ILs in related measures and excess functions are sensitive probe for the molecular interactions. Therefore, we calculated the excess molar volume (V^E) and the deviation in isentropic compressibility ($\Delta \kappa_s$) using the experimental values such as densities (ρ) and ultrasonic sound velocities (u) that are measured over the whole compositions range at five different temperatures (293.15, 298.15, 303.15, 308.15 and 313.15 K) and atmospheric pressure. These excess functions were adequately correlated by using the Redlich–Kister polynomial equation. It was observed that for all studied systems, the V^E and $\Delta \kappa_s$ values are negative for the whole composition range at 293.15 K. And, the excess function follows the sequence: 2-butanol>1-butanol>2-methyl-2-propanol, which reveals that (primary or secondary or tertiary) position of hydroxyl group influence the magnitude of interactions with ILs. The negative values of excess functions are contributions from the ion-dipole interaction, hydrogen bonding and packing efficiency between the ILs and butanol isomers. Hence, the position of hydroxyl group plays an important role in the interactions with ILs. The hydrogen bonding features between ILs and alcohols were analysed using molecular modelling program by using HyperChem 7.

INTRODUCTION

Till date numerous research groups have focused their work on the study of fascinating physical properties of ionic liquids (ILs), due to their wide variety of applications in industries and applied chemistry [1]–[16]. For the applications in chemical and industrial processes, the knowledge of the thermophysical properties of IL is essential, as they represent the basis for the chemical and biological process [17]–[28]. Many of these studies have led to new possible applications for ILs [18]–[21]. Knowledge of structure and properties of ILs is essential for the understanding of their molecular interactions in the binary mixtures [4]–[6], [8]–[16]. Nevertheless, in order to transfer the ILs from laboratory to industry, designing future processes and equipment involving these ionic compounds, an accurate knowledge about their physical properties, either for pure ILs or mixed with other solvents, is crucial. Therefore, a deep knowledge of thermophysical properties of ILs and their liquid mixtures are essentially required for scientific community. Apparently, the physicochemical properties of ILs are quite sensitive toward the structure and nature of cations and anions [7]–[11]. The variations in thermophysical properties of ILs, such as density (ρ) and speed of sound (u) are observed to be very sensitive to the change in ion, mainly due to the microscopic level interactions between solvent molecules [12]–[16].

Binary mixtures of ILs with other solvents can also improve the thermodynamic and transport properties of working fluids as well as the efficiency of the chemical equipments such as batteries, photoelectrical cells, and other electrochemical apparatus. The use of the binary mixtures of ILs with polar compounds such as alcohols allows the change and control of the properties of the mixtures to suit a given situation [29].Thermodynamic properties of mixtures containing ILs and alcohols are important for both the design of many technological processes and an understanding of the solute–solvent interactions in the mixtures. These properties are required in the development of models for process design, energy efficiency, and in the evaluation of possible environmental impacts [30]. Regarding the study of physical properties for binary mixtures of alcohol+ILs, a large number of works have been published in recent years [31]–[41], showing the interest of the scientific community for this field. While, still there is no experimental or theoretical results are available for the thermophysical properties between the tetraalkylammonium hydroxide and butanol isomers. Additionally, there is no study to show the interactions between the hydroxide anion of the IL and hydroxyl group of alcohols.

In this research to study these interaction, we explore and compare the measurements of two thermophysical properties such as ρ and u of binary mixtures involving 1-butanol, 2-butanol and 2-methyl-2-propanol with tetrapropylammonium hydroxide $[(C_3H_7)_4N][OH]$ (TPAH) and tetrabutylammonium hydroxide $[(C_4H_9)_4N][OH]$ (TBAH) ILs over a complete mole fraction range at various temperatures from 293.15 to 313.15 K, with interval of 5 K. Further, the excess molar volume (V^E), and deviation in isentropic compressibilities ($\Delta\kappa_s$) were calculated using experimental data. The resulting V^E and $\Delta\kappa_\sigma$ values were found to be strongly dependent on the place of hydroxyl group attached in the chain and also on the interactions between the hydroxide anion of ILs and hydroxyl group of the alcohols. These deviations in physical parameters have been explained in terms of intermolecular interactions between alcohols and ILs. Additionally, the temperature also plays an important role in interaction studies. Moreover, the hydrogen bonding features between ILs and alcohols were carried out to get a deep insight into intermolecular interactions for the studied compounds. These studies were performed according to the semi-empirical calculations by using HyperChem 7.

MATERIALS AND METHODS

Materials

1-butanol, 2-butanol and 2-methyl-2-propanol were obtained from Merck >99% of purity and stored over freshly activated 3 Å molecular sieves and were purified by the standard method described by Riddick et al [42]. A comparison is made for the pure alcohols in Table 1 between the experimental ρ and u values determined in the present study and those reported in the literature [31], [38], [41]–[50]. ILs were synthesized in laboratory and analysed using ¹H-NMR, the preparation is given below.

Table 1: Specifications of pure components and comparison of experimental densities (ρ) and ultrasonic sound velocities (u) with the literature values for alcohols

Solvent	ρ/(g.cm⁻³)			u/(m-s⁻¹)		
	T/K	Exptl.	Lit.	T/K	Exptl	lit.
1-Butanol	293.15	0.80980	0.80960 [ref 49]	293.15	1257	1256 [ref 42]
			080977 [ref 38]			1257 [ref 44]
			080917 [ref 46]			
			0 8094 (ref 44]			
			08098 [ref 43]			
			08095 [ref 42]			
	298.15	0.80567	080571 [ref 31]	298.15	1240	1239 [ref 42]
			0 80598 [ref 38]			1240 [ref 44]
			0 80554 [ref 46]			
			0.8058 [ref 44]			
			0 8060 [ref 43]			
			0 8057 [ref 42]			

	303.15	0.80195	080208 [ref 38]	303.15	1222	1222 [ref 42]
			0 80190 [ref 46]			1223 [ref 44]
			0 8018 [ref 44]			
			0 8021 [ref 43]			
			0.8019 (ref 42]			
			0 80221 [ref 48]			
	313 15	0 79821	0.79819 [ref 38]	308 15	1207	1206 [ref 42]
			0 7982 [ref 46]			1206 (ref 44)
			0.7979 [ref 44]			
			0.7980 [ref 43]			
			0.7980 [ref 42]			
			0.79834 [ref 48]			
	313.15	0.79420	0.79436 [ref 38]	313.15	1190	1189 [ref 42]
			0.79460 [ref 46]			
			0.7943 [ref 43]			
			0.7941 [ref 42]			
			0.79046 [ref 48]			
2-Butanel	293.15	0.80709	0.8063 [ref 42]	293.15	1229	1230 [ref 44]
			0.80657 [ref 46]			

	298 15	0.80267	0.8022 [ref 42]	298 15	1211	1212 [ref 42]
			0.80228 [ref 46]			
	303.15	0.79876	0.7980 [ref 42]	303 15	1193	1194 [ref 42]
			079892 [ref 45]			
			0.79799 [ref 46]			
			0.79835 [ref 48]			
			0.7989 [ref 47]			
	308.15	0.79446	0.7937 [ref 42]	308 15	1175	1176 [ref 42]
			0.79372 [ref 46]			
			0.79405 [ref 48]			

Solvent		p/(g.cm^{-3})		u/(m.s^{-1})		
T/K(	Exptl.	Lit.	T/K	Exptl	Lit.	
313.15	0 79007	07893 [ref 42]	313.15	1157	1158 [ref 42]	
		078943 [ref 46]				
		0.78965 (ref 48]				
		0.7901 [ref 47]				

2-Meth-yl-2-293.15 Propanol	0.78576		293.15	1145	
298.15	0.78080	0.7812 [ref 41]	298.15	1123	
303.15	0.77531	0.77531 [ref 47]	303.15	1102	
		0.77616 [ref 45]			
308.15	0.76481	0.77036 [ref 48]	308.15	1080	
313.15	0 76507	07048 [ref 47]	313.15	1059	
		0.76507 [ref 48]			

Synthesis of ILs

Synthesis of Tetrapropylammonium Hydroxide (TPAH).

The synthesis of this IL was carried out in a 250 mL round bottomed flask, which was immersed in a water-bath, fitted with a reflux condenser. Solid potassium hydroxide (40 mmol) was added to a solution of tetrapropylammonium bromide $[(C_3H_7)_4N][Br]$ (40 mmol) in dry methylene chloride (20 mL), and the mixture was stirred vigorously at room temperature for 10 h. The precipitated KBr was filtered off, and the filtrate was then evaporated to leave the crude $[(C_3H_7)_4N][OH]$ as a viscous liquid that was washed with ether (2×20 mL) and dried at 343.15 K for 5 h to obtain the pure IL. The sample was analyzed by Karl Fisher titration and revealed very low levels of water (below 70 ppm). The yield of TPAH was 82%. ^{1}H NMR (DMSOd$_6$): δ (ppm) 0.8 (t, 12H), 1.46 (m, 8H), 2.92

(t, 8H), 4.56 (s, OH). HRMS calculated for $C_{12}H_{29}NO$ (M+ - OH) 203.36, found 203.25.

Synthesis of Tetrabutylammonium Hydroxide (TBAH).

A procedure similar to that above for $[(C_3H_7)_4N][OH]$ was followed with the exception of the use of $[(C_4H_9)_4N][Br]$ ([cation]) instead of $[(C_3H_7)_4N][Br]$. The yield of TBAH was 82%. ^1H NMR (DMSOd$_6$): δ (ppm) 0.94 (t, 12H), 1.37 (m, 8H), 1.96 (m, 8H), 3.43 (t, 8H), 4.78 (s, OH). HRMS calculated for $C_{16}H_{37}NO$ (M+ - OH) 259.47 found out to be 259.34.

Experimental Procedure

Density (ρ) and speed of sound (u) measurements.

The density (*ρ*) and speed of sound (*u*) measurements were performed with an Anton-Paar DSA 5000 with an accuracy of temperature of ±0.01 K. The uncertainties in the density and speed of sound measurements were ±0.00005 g cm^{-3} and 0.01 m s^{-1} respectively. Prior to measurements, the instrument was calibrated with deionized water and dry air as standards at 293.15 K.

The binary mixtures of butanol isomers and IL were prepared by mass using a high-precision analytical balance with an uncertainty of ±1×10^{-4} g. All of the samples were prepared immediately before the measurements to avoid variations in composition due to evaporation of the solution. Clear and air bubble free solutions were used to perform the *ρ* and *u* experiments at different temperatures. The detailed measurement procedures used were described in detail in our previous research papers [12]–[15].

Hydrogen Bonding through Simulation Program.

The structures of ILs and alcohols were optimized based on molecular mechanics and semi-empirical calculations using the HyperChem 7 molecular visualization and simulation program[51]–[54]. Initial molecular geometry of butanol isomers and ILs were optimized with the PM3 semi-empirical calculations and single point calculations were carried out to determine the total energies. Now the optimized molecules, alcohols and IL were chosen and then placed on top of each other symmetrically (parallel) with a starting interplanar distance of 2.3 Å and the angle made by covalent bonds to the donor and acceptor atoms less than 120^0 was fulfilled. Further, the geometries were optimized using geometry optimizations based on molecular mechanics (using the MM+force field) and PM3 semi-empirical calculations, the Polak-Ribiere routine with rms gradient of 0.01 as the termination condition was used. PM3 uses a set of parameters derived from a variety of experimental versus calculated molecular properties, as compared to other semiempirical methods, including the AM1 procedure [53]. Typically, nonbonded interactions are less repulsive in the PM3 procedure [54]. Hydrogen bonds were displayed using HyperChem "show hydrogen bonds" and "recompute hydrogen bond" options.

RESULTS AND DISCUSSION

In order to have the better understanding of the molecular interactions between tetraalkylammonium hydroxide ILs with polar solvents such as 1-butanol, 2-butanol and 2-methyl-2-propanol, we have measured ρ and u properties over the whole composition range at various temperatures such as 293.15, 298.15, 303.15, 308.15 and 313.15 K under atmospheric pressure. The experimental ρ and u values of ILs with alcohols are presented as a function of IL concentration. Further, Figures 1–4 show the measured ρ and u values for the binary mixtures of different butanol isomers with both ILs (TPAH and TBAH) at all the studied temperatures. Figures 1 and 2 reveal that the variation of ρ values of TPAH or TBAH with 1-butanol, 2-butanol and 2-methyl-2-propanol, shows similar trends. It has been found that the ρ of the mixtures increased with the increasing concentrations of the ILs in alcohols. The effect of the ILs on the ρ in the alcohols has been examined at various temperatures. It has been observed that the ρ values decreased as temperature increased in the all systems. The results in Figure 1a clearly reveal that the ρ values of the TPAH+1-butanol mixture increase sharply up to $x1 \approx 0.8000$ and later become almost constant at all the temperatures. While for TPAH+2-butanol, ρ values increase up to very rich IL concentration $x1 \approx 0.9900$, as shown in Figure 1b. On the other hand, ρ values for TPAH+2-methyl-2-propanol increase sharply up to $x1 \approx 0.6400$ and, no prominent changes have been observed afterwards. The increase in ρ values for TPAH+alcohols mixtures is possibly due to increase in the ion pair interactions between TPAH and alcohols. This shows that the density values of the TPAH+alcohol mixtures are not affected much due to change in the position of hydroxyl group in different isomers.

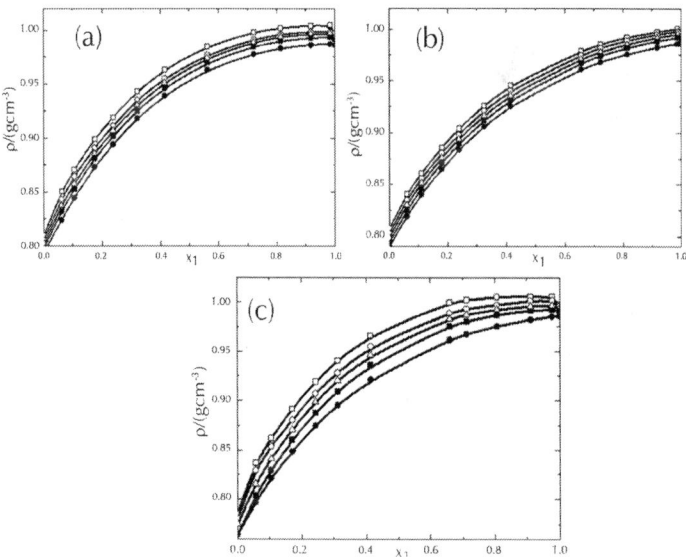

Figure 1: Densities for the mixtures of TPAH with alcohols vs mole fraction of IL$x1$ for (a) TPAH+1-butanol; (b) TPAH+2-butanol and (c) TPAH+2-methyl-2-propanol, 293.15 K (□),298.15 K (○), 303.15 K (▲), 308.15 K (■),313.15 K (●) at various compositions and at atmospheric pressure.The solid line represents the smoothness of these data.

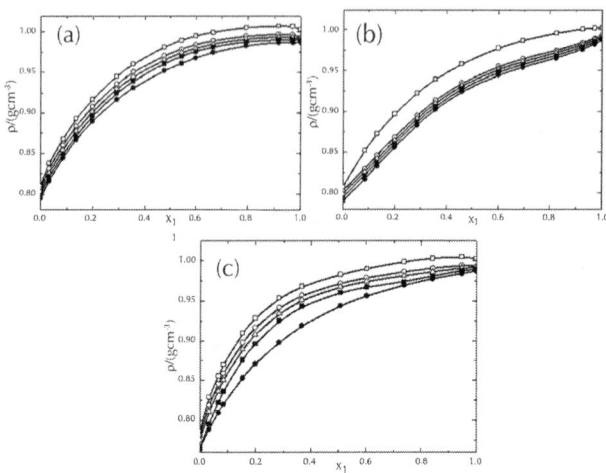

Figure 2: Densities for the mixtures of TBAH with alcohols vs mole fraction of ILx1 for (a) TBAH+1-butanol; (b) TBAH+2-butanol and (c) TBAH+2-methyl-2-propanol, 293.15 K (□), 298.15 K (○), 303.15 K (▲), 308.15 K (■),313.15 K (●) at various compositions and at atmospheric pressure The solid line represents the smoothness of these data.

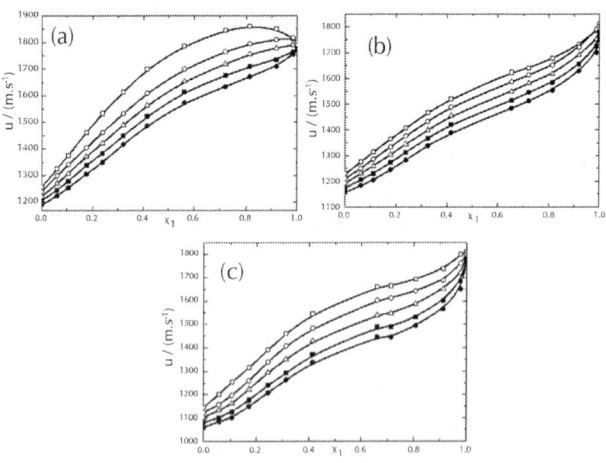

Figure 3: Ultrasonic sound velocity for the mixtures of TPAH with alcohols vs mole fraction of IL x1 for (a) TPAH+1-butanol; (b) TPAH+2-butanol and (c) TPAH+2-methyl-2-propanol, 293.15 K (□), 298.15 K (○), 303.15 K

(▲), 308.15 K (▪),313.15 K (●) at various compositions and at atmospheric pressure.The solid line represents the smoothness of these data.

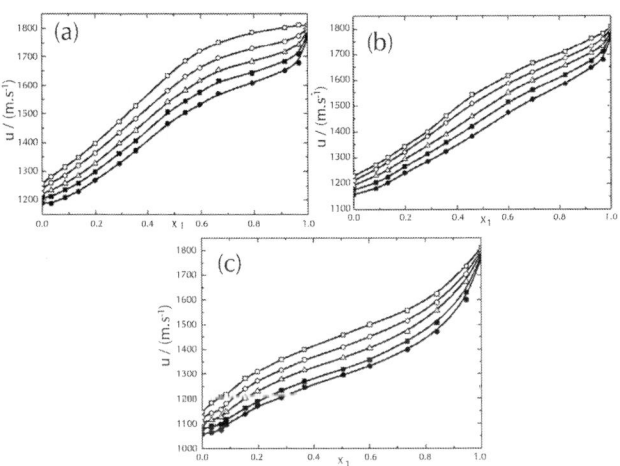

Figure 4: Ultrasonic sound velocity for the mixtures of TBAH with alcohols vs mole fraction of IL x_1 for (a) TBAH+1-butanol; (b) TBAH+2-butanol and (c) TBAH+2-methyl-2-propanol, 293.15 K (□), 298.15 K (○), 303.15 K (▲), 308.15 K (▪),313.15 K (●) at various compositions and at atmospheric pressure.The solid line represents the smoothness of these data.

Whereas, in Figure 2a the ρ values for the TBAH+1-butanol mixtures increases sharply up to$x_1 \approx 0.6200$, later the increase is marginally very less at high concentration of IL region. Whereas, TBAH+2-butanol mixtures have similar trends as shown earlier by TPAH+2-butanol,ρ values increase up to very rich IL concentration $x_1 \approx 0.9900$, as depicted by Figure 2b. Further, TBAH+2-methyl-2-propanol mixture shows the increase in ρ values, while ρ doesn›t increase sharply at mole fraction 0.5000–0.9900, which may be due to decrease in ion-pair interactions between TBAH and 2-methyl-2-propanol, as shown in Figure 2c, we observed that the densities of investigated systems increase with increasing the length of alkyl chain in IL. It was found that ρ values to be higher in the TBAH+butanol isomers as compared to TPAH+butanol isomers at equimolar mixture. Whereas, according to early documented research articles the density decreases with increase in alkyl chain in a cation or anion [55],[56]. These discrepancies vary from IL to IL and solvent to solvent and also depend on the nature as well as structural arrangement of IL and solvent. We observed that with increase in temperature, ρ values of TBAH+butanol isomers decreases more as compared to TPAH+butanol isomers. This might be due to the assumption that the ion-pair interaction decreases more for high alkyl chain+butanol isomers as compared to lower alkyl chain+butanol isomers with the increase in temperature.

Ultrasonic sound velocities (u) prove to be an informative source regarding the properties of different solvents and their mixture. The values of u were found to decrease with an increase in temperature while u values increased with increasing in mole fraction of IL. As noted fromFigures 3 and 4, there is a sharp increase of u in all ILs, except in the mixture of TPAH with 1-butanol at 293.15 K, in the mole fraction range from 0.8000 to 0.9900 of IL. Over this range, the$u$ values decrease slightly for the mixtures of TPAH with 1-butanol at 293.15 K. Whereas, no change is observed in rest of the IL+butanol isomers at all investigated temperatures. This $u$value is significantly increased in IL-solvent interactions when the mole fraction of IL was increased. If we compare the TBAH+1-butanol to TPAH+1-butanol, it has been observed that the u values slightly decrease when the alkyl substituents size of cation increases. Whereas, the same trend was observed on comparing the u values of TBAH+2-methyl-2-propanol to TPAH+2-methyl-2-propanol. It has been found that the u values slightly decrease when the size of cation increases. While the u value for TPAH+2-butanol are lower than TBAH+2-butanol, which again reveal that u values slightly increase as the size of cation increases. Hence, our results lead to conclusion that interactions of ILs with alcohols, depends upon the position of the hydroxyl group.

Thermophysical properties of mixed solvents of ILs with butanol isomers can be tunable. The extent of deviation of liquid mixtures from ideal behavior is best expressed by excess functions. Excess molar volumes (V^E) as well as ultrasonic studies are known to provide useful insights into solution structural effects and intermolecular interactions between component molecules. The extent of deviation of liquid mixtures from ideal behavior is best expressed by excess functions. Volumetric properties of binary mixtures of ILs with polar compounds are contributed to the clarification of the various intermolecular interactions existing between the different species found in solution. The excess volumes are determined from the density of pure compounds (ρ_1 and ρ_2) and mixture (ρ_m) using a standard equation [15]. The ultrasonic studies have been adequately employed in understanding the nature of molecular interaction in solvent mixed systems. In the chemical industry, knowledge of the ultrasonic and its related properties of solutions are essential in the design involving chemical separation, heat transfer, mass transfer, and fluid flow. Isentropic compressibilities (κ_s) of the binary mixtures were calculated using the relation from ρ and u. The composition dependence of the V^E and $\Delta\kappa_s$ properties represents the deviation from ideal behavior of the mixtures and provides an indication of the interactions between IL and alcohols. These properties were mathematically fitted by variable degree functions using the Redlich-Kister expression:

$$Y = x_1 x_2 \left(\sum_{i=0}^{n} a_i (x_1 - x_2)^i \right)$$

(1)

$$\sigma = \left[\frac{\sum_{i=1}^{n} \left(Y_i^{exp} - Y_i^{cal} \right)^2}{n} \right]^{1/2}$$

(2)

Where Y refers to V^E or $\Delta\kappa_s$. ai are adjustable parameters and can be obtained by least-squares analysis. The values of V^E and $\Delta\kappa_s$ for the binary mixtures at various temperatures as function of ILs concentrations. Figures 5 to10 display the experimental data for the binary mixtures, and the fitted curves, along with the excess properties of V^E and $\Delta\kappa_s$ for the butanol isomers with ILs as function of IL concentrations at different temperatures.

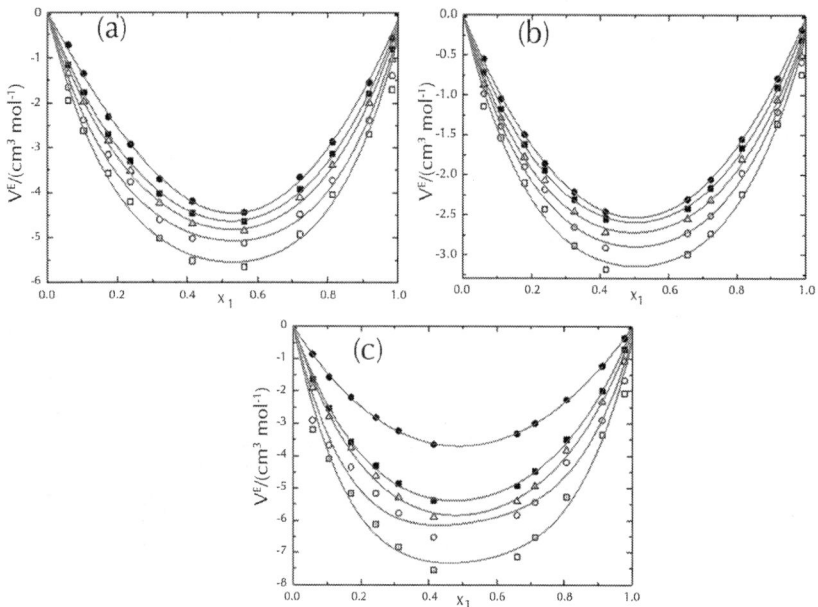

Figure 5: Excess molar volumes (V^E) against the mole fraction of TPAH $x1$ for (a) TPAH+1-butanol; (b) TPAH+2-butanol and (c) TPAH+2-methyl-2-propanol, 293.15 K (□), 298.15 K (○), 303.15 K (▲), 308.15 K (■),313.15 K (●) at various compositions and at atmospheric pressure.Solid lines correlated by the Redlich-Kister equation.

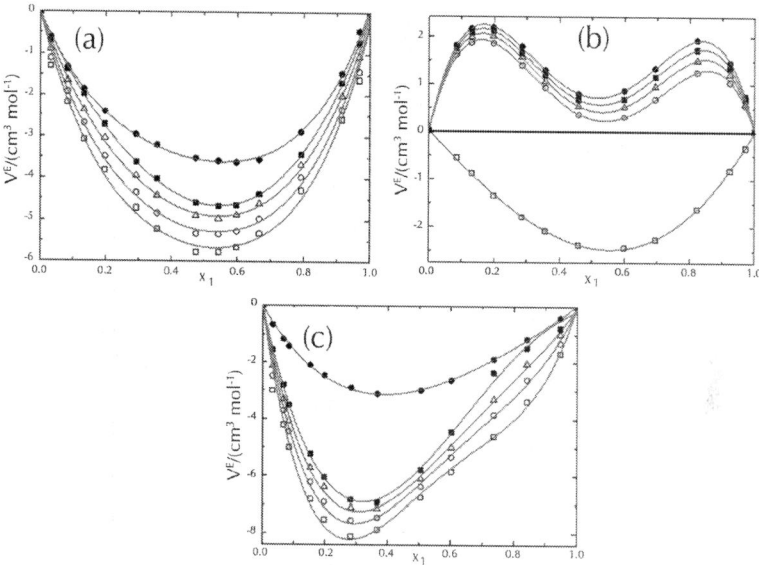

Figure 6: Excess molar volumes (V^E) against the mole fraction of TBAH $x1$ for (a) TBAH+1-butanol; (b) TBAH+2-butanol and (c) TBAH+2-methyl-2-propanol, 293.15 K (□), 298.15 K (○), 303.15 K (▲), 308.15 K (■),313.15 K (●) at various compositions and at atmospheric pressure.Solid lines correlated by the Redlich-Kister equation.

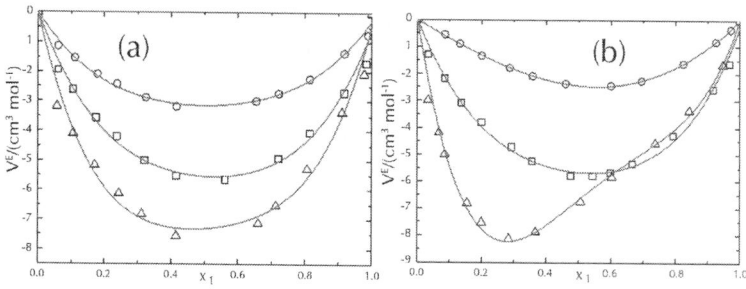

Figure 7: Excess molar volumes (V^E) of ILs+alcohols at293.15 K for (a) TPAH+1-butanol(□), TPAH+2-butanol (○) and TPAH+2-methyl-2-propanol (▲); (b) TBAH+1-butanol(□), TBAH+2-butanol (○) and TBAH+2-methyl-2-propanol (▲) at atmospheric pressure.Solid lines correlated by the Redlich-Kister equation.

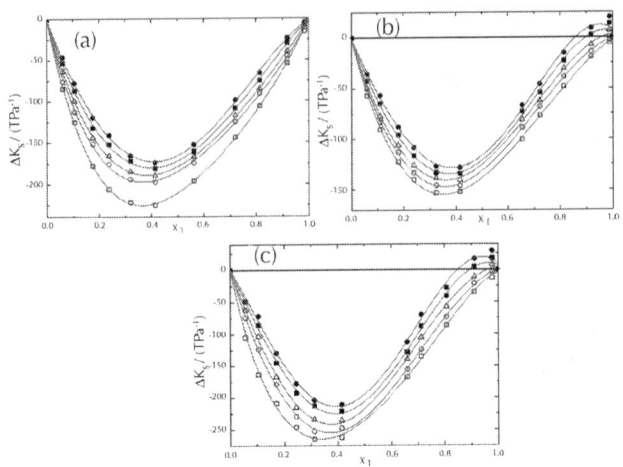

Figure 8: Deviation in isentropic compressibilities (**Δκ$_s$**) against the mole fraction of TPAH $x1$ for (a) TPAH+1-butanol; (b) TPAH+2-butanol and (c) TPAH+2-methyl-2-propanol, 293.15 K (□), 298.15 K (○), 303.15 K (▲), 308.15 K (▪),313.15 K (●) at various compositions and at atmospheric pressure.Solid lines correlated by the Redlich-Kister equation.

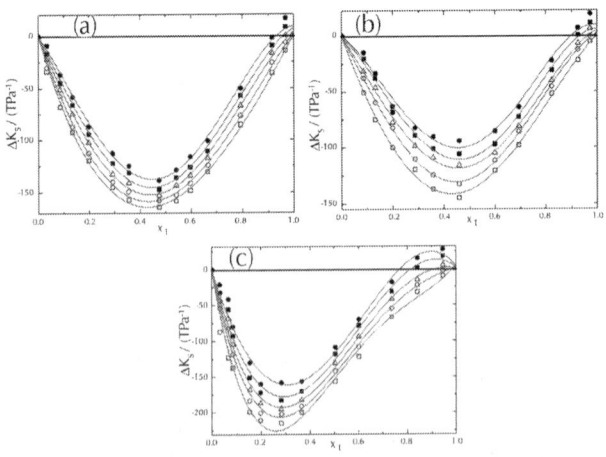

Figure 9: Deviation in isentropic compressibilities (**Δκ$_s$**)against the mole fraction of TBAH $x1$ for (a) (a) TBAH+1-butanol; (b) TBAH+2-

butanol and (c) TBAH+2-methyl-2-propanol, 293.15 K (□), 298.15 K (○), 303.15 K (▲), 308.15 K (■),313.15 K (●) at various compositions and at atmospheric pressure.Solid lines correlated by the Redlich-Kister equation.

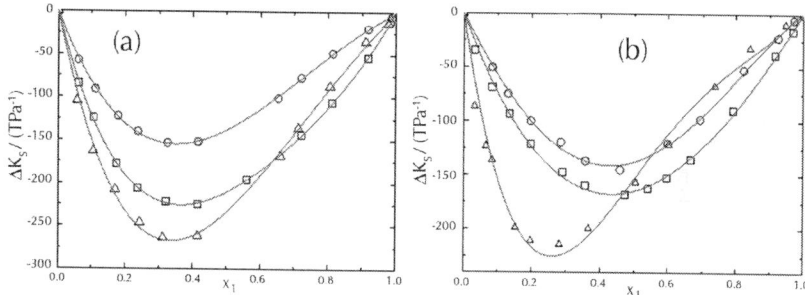

Figure 10: Deviation in isentropic compressibilities ($\Delta\kappa_s$) of ILs+alcoholsx1 at293.15 K for (a) TPAH+1-butanol(□), TPAH+2-butanol (○)and TPAH+2-methyl-2-propanol(▲);(b) TBAH+1-butanol(□), TBAH+2-butanol (○) and TBAH+2-methyl-2-propanol(▲)at atmospheric pressure.Solid lines correlated by the Redlich-Kister equation.

From Figure 5 one can note that the values of V^E are negative for all TPAH+butanol isomers systems at all measured temperatures over whole composition range. We have observed that the excess molar volumes present a minimum at $x1\approx0.5634$ for TPAH+1-butanol at all investigated temperatures, whereas we obtained that the V^E values present a minimum at $x1\approx0.4143$ for the TPAH+2-butanol system. Further, minimum V^E values lie at $x1\approx0.4244$ for TPAH+2-methyl-2-propanol system at all investigated temperatures. The minimum V^E values could be due to hydrogen bonds between alcohols and TPAH IL. The decrease in the magnitude of the negative V^E values with an increase in the IL composition can be attributed to the decrease of hydrogen bonding. In other words, due to increase in the concentration of the IL results in decrease of packing efficiency. Further, with increase in temperature the magnitude of the negative V^E values decreases in all the TPAH+butanol isomers systems. This can again be due to decrease in magnitude of the hydrogen bonding with increase in the temperature.

It is interesting to note that the V^E values in TPAH+2-methyl-2-propanol mixture show more negative values of V^E at the alcohol-rich composition than the TPAH+1-butanol and TPAH+2-butanol mixtures at 293.15 K (Figure 7a), implying that in the TPAH+2-methyl-2-propanol, there are ion-dipole interactions and packing effects with 2-methyl-2-propanol which are stronger than those in the 2-butanol and 1-butanol solution at $x1 \approx 0.4200$. A comparison between the negative deviation of V^E of TPAH+2-methyl-2-propanol, TPAH+1-butanol and TPAH+2-butanol suggests that there is a difference of the hydroxyl position in the alkyl chain, leading to variation in the interactions between the alcohols and TPAH. Molecular interaction between TPAH and alcohols follows the following order at 293.15 K, 2-methyl-2-propanol>1-butanol>2-butanol.

Further, the negative V^E values are observed for TBAH+butanol isomers at all measured temperatures over whole composition range, except 2-butanol at higher temperatures 293.15 to 313.15 K. The V^E values for 1-butanol with TBAH are as represented in Figure 6a. And, we found that negative V^E values are observed over the entire mole fraction range at all investigated temperatures. These negative V^E values reveal that a more efficient packing or attractive interaction occurred between the TBAH and 1-butanol. 1-Butanol forms a hydrogen bond with the alkyl chain cation, while the interactions decrease at higher temperatures. The interactions between the 1-butanol molecules and the alkyl chain of TBAH are due to ion-dipole or hydrogen bonding interactions. This will reduce the interactions between the tetrabutylammonium cation and hydroxide anion in the IL, which contributes to the negative V^E values. Furthermore, the observed positive V^E values for TBAH+2-butanol at higher temperatures show that there exist no specific interactions between unlike molecules, as displayed in Figure 6b. The magnitude and sign of V^E values are a reflection of the type of interaction staking place in the mixture, which are the result of different effects containing the loss of the dipole interaction from each other and the breakdown of the IL ion pair (positive V^E). The interaction between the ion pair of ILs increases as compared to IL+2-butanol interactions, which leads to positive contribution at higher temperature over whole composition range. While,Figure 6c shows the negative V^E values for TBAH+2-methyl-2-propanol at all measured temperatures over whole composition range. This might be due to the large difference between the molar volumes of the 2-methyl-2-propanol and TPAH implying that it is possibly due to the fact that the relatively small organic molecules fit into the interstices upon mixing. Therefore, the filling effect of organic molecular liquids in the interstices of ILs, and the ion-dipole interactions between organic molecular liquid and alkyl cation of ammonium ILs, all contribute to the negative values of V^E.

Clearly, the observed negative V^E values increases further with increasing the temperature in the entire mole fraction range for all IL systems. It is interesting to note that the V^E values in the ILs+2-methyl-2-propanol mixture shows more negative values of V^E than the IL+2-butanol and IL+1-butanol mixtures at 298.15 K over the alcohol rich concentration range (Figure 7), implying that in the 2-methyl-2-propanol there is strong ion-dipole interactions and packing effects with ILs as compared to 2-butanol and 1-butanol. The magnitude and sign of V^E values are a reflection of the type of interactions taking place in the mixture, which reveals that V^E values are more negative for TBAH ($V^E = -8.149$ cm^3.mol^{-1} at $x_1 = 0.2831$ for TBAH+2-methyl-2-propanoland $V^E = -5.787$ cm^3.mol^{-1} at $x_1 = 0.5428$ for TBAH+2-butanol) than TPAH ($V^E = -7.547$ cm^3.mol^{-1} at $x_1 = 0.4143$ for TPAH+2-methyl-2-propanol and $V^E = -5.702$ cm^3.mol^{-1} at $x_1 = 0.5634$ for TPAH+2-butanol) in all systems except in ILs+1-butanol systems. For TBAH+1-butanols ($V^E = -2.448$ cm^3.mol^{-1} at $x_1 = 0.5993$) and $V^E = -3.175$ cm^3.mol^{-1} at $x_1 = 0.4139$ for TPAH+1-butanol, hence the V^E is more negative for the TPAH+1-butanol than TBAH+1-butanol due to steric hindrance created by the long chain cation of TBAH, that reduces the interaction magnitude between TBAH and 1-butanol. Interestingly, the hydrogen bonding between ILs and butanol isomers has predicted using semiempirical calculations with the help of Hyperchem 7, and those interactions are explicitly elucidated. Using semiempirical calculations for the hydrogen bonding between ILs and butanol isomers, we calculated heat of formation of the complexes and compared the values with those of the ILs and butanol isomers (as displayed in Table 2). In all the cases, ΔH_f of the complex resulting from hydrogen bonding was higher than the sum of ΔH_f's of butanol isomers and ILs. It is reasonable to assume that these differences ($\Delta\Delta H_f$), calculated according to equation 3, represent the energies of the hydrogen bond. The energies of the hydrogen bonding can also obtained by using the total binding energies of butanol isomers [54] and ILs, presented in Table 2, instead of ΔH_f's for these calculations. The results in Table 2 indicate that the energy required for the formation of a weak hydrogen bond is less than required for the formation of a stronger hydrogen bond:

$$\Delta\Delta H_f = \Delta H_f(1) + \Delta H_f(2) - \Delta H_f(3)$$

(3)

Where ΔH_f (1) is the heat of formation of the butanol isomers, ΔH_f (2) the heat of formation of the ILs and ΔH_f (3) the heat of formation of the complex (butanol isomers and ILs).

Table 2: Calculated binding energies (E), heats of formations (ΔH_f), and estimated hydrogen bond energies ($\Delta\Delta H_f$) (kcal/mol)

Solvent	E/(kcal/mol)	ΔH_f (kcal/mol)	$\Delta\Delta H_f$ (kcal/mol)
TPAH	—3784.96	—50.76	
TBAH	—4907.84	—73.27	
1-Butanol	—1330.63	—66.49	
2-Butanol	—1330.86	—66.72	
2-Methyl-2-Propanol	—1333.75	—69.61	

TPAH+1-Bu-tanol	—5117.47	—119.12	1.87
TPAH+2-Bu-tanol	—5119.10	—121.62	4.14
TPAH+2-Methy1-2-Pro-panol	—5118.74	—121.01	0.64
TBAH-F1-Butanol	—6238.88	—140.15	0.39
TBAH+2-Bu-tanol	—6241.01	—142.28	2.29
TBAH+2-Methy1-2-Pro-panol	—6241.09	—142.96	0.08

The hydrogen bonding between nitrogen group of TPAH IL with the "-OH" group of 1-buatnol. The binding energy of the TPAH is found to be −3784.96 kcal/mol and that of 1-butanol is found to be −1330.63 kcal/mol, but after the hydrogen bonding occur between the TPAH and 1-butanol, the binding energy of complete system comes out to be −5117.47 kcal/mol (Table 2). Hence, the estimated hydrogen bond energy of the above system is ≈1.87 kcal/mol, which could probably due to the interaction of TPAH with 1-butanol that leads to decrease in energy (less than sum of individual energy of TPAH and 1-butanol), and increase in the strength of hydrogen bonding. Depict the possibility of hydrogen bonding between the nitrogen group of TPAH IL with hydroxyl group of 2-butanol and 2-methyl-2-propanol and the estimated hydrogen bond energies are ≈4.14 and ≈0.64 kcal/mol respectively (Table 2). This shows that the strength of hydrogen bonding is more for TPAH+2-butanol as compared to other butanol isomers. clearly again show the hydrogen bonding between the nitrogen group of TBAH with hydroxyl group of 1-butanol, 2-butanol and 2-methyl-2-propanol and now the hydrogen bond energies are ≈0.39, ≈2.29 and ≈0.08 respectively. Hence, we may conclude that the hydrogen bond in case of TBAH+2-butanol is stronger as compared to other butanol isomers.

Our interpretation of hydrogen bonding between of IL and butanol isomers (based V^E data) is quite corroborated with our theoretical calculation of hydrogen bonding of IL+butanol isomers. It is noteworthy that the hydroxyl groups of alcohols are interacting with the nitrogen group of ILs (TPAH and TBAH) According to literature, the negative V^Evalues are a result of contributions from both the accommodation of organic molecules in the interstice of the IL networks and the ion–dipole interactions between the organic molecules and cation of the ionic liquid [38], [57]. Our experimental results reveal that the negative V^E values for entire composition and theoretical calculation suggested that hydroxyl groups of alcohols are interacting with the cation of ILs, Hence, our results are very well correlated with literature results. Therefore, the absolute value of V^E is an indicative to the difference in the packing efficiency and the interaction intensity. As can be seen from Figure 7, the V^E values for the studied systems follow the sequence: 2-methyl-2-propanol>1-butanol>2-butanol. If only ion–dipole interactions are taken into consideration, the order 1-butanol>2-butanol>2-methyl-2-propanol is understandable. The decreased dielectric constant from 1-butanol (17.8), 2-butanol (16.6) and 2-methyl-2-propanol (10.9) leads to the weaker ion–dipole interaction and in turn resulting in the smaller V^E values. Whereas, if we consider the energies of the hydrogen bond (Table 2), the order for ILs+butanol isomers follows: 2-butanol>1-butanol>2-methyl-2-propanol. Our experimental results suggest the order 2-methyl-2-propanol>1-butanol>2-butanol, which reveal that the interactions are not only due to individual contribution of ion-dipole interaction or H-bonding, but it is the combined effect of both the factors. Whereas, another plausible reason is that the butanol isomers makes it easy to accommodate in the interstice of the IL network, and the higher packing efficiency also leads to the larger V^E values. While, with increase in temperature there is decrease in V^E values in all the systems because at higher temperature the packing efficiency decreases of ILs. On the other hand, the ion-dipole interactions also decrease with the increase in temperature that leads to decrease in V^E values.

Further, for better understanding of the interactions between the tetraalkylammonium hydroxide ILs, we have calculated the $\Delta\kappa_s$. As seen in Figure 8, $\Delta\kappa_s$ values of tetraalkylammonium hydroxide ILs+butanol isomers are negative over the full composition range at 293.15 K as a function of ILs concentration. The behavior of $\Delta\kappa_s$, implies that these mixtures are less compressible than the ideal mixture. This is due to closer approach of unlike molecules and a stronger interaction between components of mixtures that leads to a decrease in the compressibility. From Figure 8a, it can be seen that the minimum $\Delta\kappa_s$ values are observed at mole fraction of IL ≈ 0.4141 for the TPAH+1-butanol system. The negative $\Delta\kappa_s$ values of TPAH+1-butanol are attributed to the strong attractive interactions due to the solvation of the ions in these solvents, over the complete composition range and at all studied temperatures. Similarly, the curves in Figure 8 (c and d), show that the $\Delta\kappa_s$ values for the 2-butanol or 2-methyl-2-propanol+TPAH systems are negative over the complete composition range and at all studied temperatures, except at the higher temperatures (308.15 and 313.15 K) for ≈ 0.8100 to 0.9999 composition range. The minimum is approached at mole fraction of IL ≈ 0.4171, ≈ 0.3272 and ≈ 0.3138 for the TPAH+1-butanol, TPAH+2-butanol and TPAH+2-methyl-2-propanol systems at all temperatures, respectively. The negative $\Delta\kappa_s$ values attributed to the strong attractive interactions between the molecules of the components. The negative values of $\Delta\kappa_s$ of the TPAH with butanol isomers imply that solvent molecules around solute are less compressible than the solvent molecules in the bulk solutions. Whereas on further addition of IL, there is decrease in the compressibility graph at all studied temperature ranges. This might be due to the decreased attraction between IL and butanol isomers in IL rich concentration region. Additionally, for the 2-butanol and 2-methyl-2-propanol, there are positive $\Delta\kappa_\sigma$ values at higher temperatures, this is might be again due to decrease in attraction of TPAH and alcohol molecules in the IL-rich concentration region, since the interaction between the ILs increases and whereas decreases in case of IL and alcohols.

Figure 9 depicts the negative $\Delta\kappa_s$ values of all TBAH+butanol isomers over the full composition range at 293.15 K. The curves in Figure 9 show that the $\Delta\kappa_s$ values for the 1-butanol or 2-butanol or 2-methyl-2-propanol systems are negative over the complete composition range at low temperature. The minimum is approached at mole fractions of IL ≈0.4787, ≈0.4604 and ≈0.1985 for the TBAH+1-butanol, TBAH+2-butanol and TBAH+2-methyl-2-propanol systems, respectively. Our results show that for all the system, TBAH+butanol isomers shows the positive $\Delta\kappa_s$ values in the IL-rich region at the higher temperatures. These results are very similar with the TPAH+alcohols at higher temperature, which might be due to the decrease in the attraction of TBAH and alcohol molecules in the IL-rich concentration region due to the increased interaction between the ILs and the decreased interaction between IL and alcohols. Obviously, the $\Delta\kappa_s$ values in the ILs+2-methyl-2-propanol mixture shows more negative values of $\Delta\kappa_s$ than the ILs+2-butanol and ILs+1-butanol mixtures at 293.15 K over the entire concentration range (Figure 10), implying that in the 2-methyl-2-propanol there is strong ion-dipole interactions and packing effects with ILs as compared to 2-butanol and 1-butanol. The magnitude and sign of $\Delta\kappa_s$ values are a reflection of the type of interactions taking place in the mixture, which reveals that $\Delta\kappa_s$ values are more negative for TBAH+2-methyl-2-propanol ($\Delta\kappa_s$ = −211.351TPa⁻¹ at x_1 = 0.1985), than TBAH+1-butanol ($\Delta\kappa_s$ = −168.519TPa⁻¹ at x_1 = 0.4787), and least is for TBAH+2-butanol ($\Delta\kappa_s$ = −144.998TPa⁻¹ at x_1 = 0.4604). For TPAH+2-methyl-2-propanol ($\Delta\kappa_s$ = −264.189TPa⁻¹ at x_1 = 0.3115), TPAH+1-butanol ($\Delta\kappa_s$ = −222.322TPa⁻¹ at x_1= 0.3218), and TPAH+2-butanol ($\Delta\kappa_s$ = −153.949TPa⁻¹ at x_1 = 0.0.3244). TPAH+butanol isomers have more negative $\Delta\kappa_s$ values then TBAH+butanol isomers this is might be due to the steric hindrance created by the long chain cation of TBAH, which reduces the interaction magnitude between TBAH and alcohols.

However, after close look about the physical properties of alcohols with ILs, we observed that hydroxyl position of alcohols are playing important role in addition to the cation chain length of the ILs. ILs (TPAH or TBAH) interact strongly with the 2-methyl-2-

propanol as compared to the 2-butanol and 1-butanol, this is might be due to more+I-effect of 2-methyl-2-propanol, which increases its tendency to interact with ILs more strongly as compared to 2-butanol and 1-butanol. Also, 1-butanol interacts more strongly as compared to 2-butanol with ILs, which might be due to the steric hindrance the interaction of 2-butanol decreases as compared to 1-butanol. However, there is decrease in V^E and $\Delta\kappa_s$ values in all the system due to increase in temperature; this might be due to strong self-association between the alcohol molecules that prevents the alcohol-IL strong interactions. Our experimental results of V^E values are very well supported with literature [38], [46], [47]. Wen-Lu Weng [47], showed the interactions of anisole with 2-butanol and 2-methyl-2-propanol, author observed that V^E values of the 2-methyl-2-propanol is more negative than 2-butanol. Additionally, the V^E values increases (less negative) with increase in temperature. Further, Qian et al. [38], showed that 1-methylimidazolium acetate IL interacts with methanol, ethanol, 1-propanol and 1-butanol, it was observed that in all the systems V^E values increase with increase in the temperature. Moreover, during the interaction of formamide with 1-butanol negative V^E values have been observed, whereas interaction of formamide with 2-butanol results in positive V^E values [46]. These all results by various authors support our above results explanation that interactions between ILs+alcohols depend upon the position of hydroxyl group. Therefore, the physicochemical properties of ILs are quite sensitive toward the structure and nature of interacting molecules.

CONCLUSIONS

We have performed and compared thermophysical properties of binary mixtures of tetraalkylammonium hydroxide based ILs with butanol isomers over the whole composition range at various temperatures (293.15 to 313.15 K, in steps of 5 K). To obtain a more detailed picture of the molecular interactions, we measured temperature dependence properties of and u for ILs with butanol isomers over the whole composition range at various temperatures.

The and u values increase with the increasing the cation alkyl chain length of ILs. Our results reveal that the position of hydroxyl group in alcohols leads to alteration of the thermophysical properties of ILs. To measure the non-ideality of the mixtures, we determined V^E and $_s$ at each temperature as a function of IL concentration. The predicted properties were correlated by the Redlich-Kister type equation. Our studies demonstrate that there is decrease in V^E and $_s$ values in all the systems due to increase in temperature; this might be due to strong self-association between the alcohol molecules that prevents the alcohol-IL strong interactions. Additionally, according to the theoretical calculations obtained by HyperChem 7, the energy of hydrogen bond is more for low alkyl chain ILs (TPAH) as compared to higher alkyl chain ILs (TBAH) with alcohols. Molecular interactions such as ion-dipole and hydrogen bonding between the butanol isomers and alkyl chain of ILs are suggested to be mainly responsible for variation in the thermophysical parameters. Our findings provide better molecular interactions for the mixing of the solvents and better analysis of the solvation process.

AUTHOR CONTRIBUTIONS

Conceived and designed the experiments: PA. Performed the experiments: PA. Analyzed the data: PA PV. Contributed reagents/materials/analysis tools: KYB ITK EHC. Wrote the paper: PA.

REFERENCES

1. Janikowski J, Razali MR, Forsyth CM, Nairn KM, Batten SR, et al. (2013) Physical Properties and Structural Characterization of Ionic Liquids and Solid Electrolytes Utilizing the Carbamoyl cyano(nitroso) methanide Anion. Chem Plus Chem 78: 486–497. doi: 10.1002/cplu.201300068

2. Capelo SB, Méndez-Morales T, Carrete J, Lago EL, Vila J, et al. (2012) Effect of Temperature and Cationic Chain Length on the Physical Properties of Ammonium Nitrate-Based

Protic Ionic Liquids. J Phys Chem B 116: 11302–11312. doi: 10.1021/jp3066822

3.   Liu ZP, Wu XP, Wang WC (2006) A Novel United-Atom Force Field for Imidazolium-Based Ionic Liquids. Phys Chem Chem Phys 8: 1096–1104. doi: 10.1039/b515905a

4.   Chen T, Chidambaram M, Liu Z, Berend SB, Bell AT (2010) Viscosities of the Mixtures of 1-Ethyl-3-Methylimidazolium Chloride with Water, Acetonitrile and Glucose: A Molecular Dynamics Simulation and Experimental Study. J Phys Chem B 114: 5790–5794. doi: 10.1021/jp911372j

5.   Couadou E, Jacquemin J, Galiano H, Hardacre C, Anouti MA (2013) Comparative Study on the Thermophysical Properties for Two Bis[(trifluoromethyl) sulfonyl] imide-Based Ionic Liquids Containing the Trimethyl-Sulfonium or the Trimethyl-Ammonium Cation in Molecular Solvents. J Phys Chem B 117: 1389–1402. doi: 10.1021/jp308139r

6.   Fox ET, Paillard E, Borodin O, Henderson WA (2013) Physicochemical Properties of Binary Ionic Liquid–Aprotic Solvent Electrolyte Mixtures. J Phys Chem C 117: 78–84. doi: 10.1021/jp3089403

7.   Zhang S, Sun N, He X, Lu X, Zhang X (2006) Physical Properties of Ionic Liquids: Database and Evaluation. J Phys Chem Ref Data 35: 1475–1517. doi: 10.1063/1.2204959

8.   González B, Calvar N, González E, Domínguez A (2008) Density and Viscosity Experimental Data of the Ternary Mixtures 1-Propanol or 2-Propanol+Water+1-Ethyl-3-MethylimidazoliumEthylsulfate. Correlation and Prediction of Physical Properties of the Ternary Systems. J Chem Eng Data 53: 881–887. doi: 10.1021/je700700f

9.   Pereiro AB, Legido JL, Rodríguez A (2007) Physical Properties Of Ionic Liquids Based on 1-Alkyl-3-Methylimidazolium Cation and Hexafluorophosphate as Anion and Temperature Dependence. J Chem Thermodynamics 39: 1168–1175. doi: 10.1016/j.jct.2006.12.005

10. Annat G, Forsyth M, MacFarlane DR (2012) Ionic Liquid Mixtures Variations in Physical Properties and Their Origins

in Molecular Structure. J Phys Chem B 116: 8251–8258. doi: 10.1021/jp3012602

11.   Deetlefs M, Seddon KR, Shara M (2006) Predicting Physical Properties of Ionic Liquids. Phys Chem Chem Phys 8: 642–649. doi: 10.1039/b513453f

12.   Kavitha T, Attri P, Venkatesu P, Ramadevi RS, Hofman T (2012) Influence of Alkyl Chain Length and Temperature on Thermophysical Properties of Ammonium-Based Ionic Liquids with Molecular Solvent. J Phys Chem B 116: 4561–4574. doi: 10.1021/jp3015386

13.   Attri P, Venkatesu P, Hofman T (2011) Temperature Dependence Measurements and Structural Characterization of Trimethyl Ammonium Ionic Liquids with a Highly Polar Solvent. J Phys Chem B 115: 10086–10097. doi: 10.1021/jp2059084

14.   Attri P, Venkatesu P, Kumar A (2010) Temperature Effect on the Molecular Interactions Between Ammonium Ionic Liquids and N,N-Dimethylformamide. J Phys Chem B 114: 13415–13425. doi: 10.1021/jp108003x

15.   Attri P, Reddy PM, Venkatesu P, Kumar A, Hofman T (2010) Measurements and Molecular Interactions for N,N-Dimethylformamide with Ionic Liquid Mixed Solvents. J Phys Chem B 114: 6126–6133. doi: 10.1021/jp101209j

16.   Hou M, Xu Y, Han Y, Chen B, Zhang W, et al. (2013) Thermodynamic Properties of Aqueous Solutions of Two Ammonium-Based Protic Ionic Liquids at 298.15 K. J Mol Liqs 178: 149–155. doi: 10.1016/j.molliq.2012.11.030

17.   Attri P, Venkatesu P, Kumar A (2011) Activity and Stability of -Chymotrypsin in Biocompatible Ionic Liquids: Enzyme Refolding by Triethyl Ammonium Acetate. Phys Chem Chem Phys 13: 2788–2796. doi: 10.1039/c0cp01291b

18.   Seddon KR (1997) Ionic Liquids for Clean Technology. J Chem Technol Biotechnol 68: 351–356. doi: 10.1002/(sici)1097-4660(199704)68:4<351::aid-jctb613>3.0.co;2-4

19. Greaves TL, Drummond C (2008) Protic Ionic Liquids: Properties and Applications. J Chem Rev 108: 206–237. doi: 10.1021/cr068040u

20. Rogers RD, Seddon KR (2003) Ionic Liquids-Solvents of the Future?. Science 302: 792–793. doi: 10.1126/science.1090313

21. Davis JH (2004) Task-Specific Ionic Liquids. Chem Lett 33: 1072–1077. doi: 10.1246/cl.2004.1072

22. Attri P, Venkatesu P, Kumar A (2011) Activity and Stability of -Chymotrypsin in Biocompatible Ionic Liquids: Enzyme Refolding by Triethyl Ammonium Acetate. Phys Chem Chem Phys 13: 2788–2796. doi: 10.1039/c0cp01291b

23. Chiappe C, Pomelli CS, Rajamani S (2011) Influence of Structural Variations in Cationic and Anionic Moieties on the Polarity of Ionic Liquids. J Phys Chem B 115: 9653–9661. doi: 10.1021/jp2045788

24. Attri P, Venkatesu P, Kumar A (2012) Water and a Protic Ionic Liquid Acted as Refolding Additives for Chemically Denatured Enzymes. Org Biomol Chem 10: 7475–7478. doi: 10.1039/c2ob26001h

25. Noda A, Bin Hasan Susan A, Kudo K, Mitsushima S, Hayamizu K, et al. (2003) Brønsted Acid–Base Ionic Liquids as Proton-Conducting Nonaqueous Electrolytes. J Phys Chem B 107: 4024–4033. doi: 10.1021/jp022347p

26. Attri P, Lee SH, Hwang SW, IL Kim J, Lee SW, et al. (2013) Elucidating Interactions and Conductivity of Newly Synthesised Low Bandgap Polymer with Protic and Aprotic Ionic Liquids. PloS One 2013, 88: e68970 doi:10.1371/journal.pone.0068970.

27. Attri P, Choi EH (2013) Influence of Reactive Oxygen Species on the Enzyme Stability and Activity in the Presence of Ionic Liquids. PloS One 2013, 8: e75096 doi:10.1371/journal.pone.0075096.

28. Reddy PM, Venkatesu P (2011) Ionic Liquid Modifies the Lower Critical Solution Temperature (LCST) of Poly(N-

isopropylacrylamide) in Aqueous Solution. J Phys Chem B 115: 4752–4757. doi: 10.1021/jp201826v

29. Abareshi M, Goharshadi EK, Zebarjad SM (2009) Thermodynamic Properties of the Mixtures of Some Ionic Liquids with Alcohols using a Simple Equation of State. J Mole Liq 149: 66–73. doi: 10.1016/j.molliq.2009.08.004

30. Arce A, Rodil E, Soto A (2006) Physical and Excess Properties for Binary Mixtures of 1-Methyl-3-Octylimidazolium Tetrafluoroborate, [Omim][BF4], Ionic Liquid with Different Alcohols. J Solution Chem 35: 63–78. doi: 10.1007/0953-006-8939-y

31. González EJ, González B, Macedo EA (2013) Thermophysical Properties of the Pure Ionic Liquid 1-Butyl-1-methylpyrrolidinium Dicyanamide and Its Binary Mixtures with Alcohols. J Chem Eng Data 58: 1440–1448. doi: 10.1021/je300384g

32. González EJ, González B, Calvar N, Domínguez A (2007) Physical Properties of Binary Mixtures of the Ionic Liquid 1-Ethyl-3-Methylimidazolium Ethylsulfate with Several Alcohols at T = (298.15, 313.15 K, and 328.15) K and Atmospheric Pressure. J Chem Eng Data 52: 1641–1648. doi: 10.1021/je700029q

33. Lehmann J, Rausch MH, Leipertz A, Fröba AP (2010) Density and Excess Molar Volumes for Binary Mixtures of Ionic Liquid 1-Ethyl-3-Methylimidazolium Ethylsulfate with Solvents. J Chem Eng Data 55: 4068–4074. doi: 10.1021/je1002237

34. Vercher E, Orchillés AV, Miguel PJ, Martínez-Andreu A (2007) Volumetric and Ultrasonic Studies of 1-Ethyl-3-Methylimidazolium Trifluoromethane sulfonate Ionic Liquid with Methanol, Ethanol, 1- Propanol, and Water at Several Temperatures. J Chem Eng Data 52: 1468–1482. doi: 10.1021/je7001804

35. Rilo E, Ferreira AGM, Fonseca IMA, Cabeza O (2010) Densities and Derived Thermodynamic Properties of Ternary Mixtures 1-Butyl-3-Methyl-Imidazolium

Tetrafluoroborate+Ethanol+Water at Seven Pressures and Two Temperatures. Fluid Phase Equilib 296: 53–59. doi: 10.1016/j. fluid.2010.03.039

36. Anouti M, Jacquemin J, Lemordant D (2010) Volumetric Properties, Viscosities, and Isobaric Heat Capacities of Imidazolium Octanoate Protic Ionic Liquid in Molecular Solvents. J Chem Eng Data 55: 5719–5728. doi: 10.1021/ je100671v

37. Alvarez VH, Mattedi S, Martin-Pastor M, Aznar M, Iglesias M (2011) Thermophysical Properties of Binary Mixtures of {Ionic Liquid 2-Hydroxy Ethylammonium Acetate+(Water, Methanol, or Ethanol)}. J Chem Thermodyn 43: 997–1010. doi: 10.1016/j.jct.2011.01.014

38. Qian W, Xu Y, Zhu H, Yu C (2012) Properties of Pure 1-Methylimidazolium Acetate Ionic Liquid and its Binary Mixtures with Alcohols. J Chem Thermodyn 49: 87–94. doi: 10.1016/j.jct.2012.01.013

39. Gómez E, Calvar N, Macedo EA, Domínguez A (2012) Effect of the Temperature on the physical Properties of Pure 1-Propyl-3-MethylimidazoliumBis(Trifluoromethylsulfonyl)Imideand Characterization of its Binary Mixtures with Alcohols. J Chem Thermodyn 45: 9–15. doi: 10.1016/j.jct.2011.08.028

40. Jiang H, Wang J, Zhao F, Qi G, Hu Y (2012) Volumetric and Surface Properties of Pure Ionic Liquid n-Octylpyridinium Nitrate and its Binary Mixtures with Alcohols. J Chem Thermodyn 47: 203–208. doi: 10.1016/j.jct.2011.10.013

41. Domanska U, Zawadzki M, Lewandrowska A (2012) Effect of Temperature and Composition on the Density, Viscosity, Surface Tension, and Thermodynamic Properties of Binary Mixtures of N-OctylisoquinoliniumBis(Trifluoromethyl sulfonyl)Imidewith Alcohols. J Chem Thermodyn 48: 101–111. doi: 10.1016/j.jct.2011.12.003

42. Riddick JA, Bunger WB, Sakano TK (1986) Organic Solvents 4th ed., Wiley-Inter Science, New York.

43. Mokhtarani B, Sharifi A, Mortaheb HR, Mirzaei M, Mafi M, et al. (2009) Density and Viscosity of 1-Butyl-3-Methylimidazolium Nitrate with Ethanol, 1-Propanol, or 1-Butanol at Several Temperatures. J Chem Thermodyn 41: 1432–1438. doi: 10.1016/j.jct.2009.06.023

44. De Cominges BE, Piñeiro MM, Mosteiro L, Mascato1 E, Mato MM, et al. (2002) Temperature Dependence of Thermophysical Properties of Octane+1-Butanol System. J Therm Anal Calorim 70: 217–227.

45. Gnanakumari P, Venkatesu P, Mohan KR, Rao MVP, Prasad DHL (2007) Excess Volumes and Excess Enthalpies of N-Methyl-2-Pyrrolidone with Branched Alcohols. Fluid Phase Equilibria 252: 137–142. doi: 10.1016/j.fluid.2006.12.012

46. Nain AK (2007) Densities and Volumetric Properties of Binary Mixtures of Formamide with 1-Butanol, 2-Butanol, 1,3-Butanediol and 1,4-Butanediol at Temperatures between 293.15 and 318.15 K. J Solution Chem 36: 497–516. doi: 10.1007/0953-007-9122-9

47. Weng WL (1999) Densities and Viscosities for Binary Mixtures of Anisole with 2-Butanol, 2-Methyl-1-propanol, and 2-Methyl-2-propanol. J Chem Eng Data 44: 788–791. doi: 10.1021/je980283z

48. Bravo-Sanchez MG, Iglesias-Silva GA, Estrada-Baltazar A (2010) Densities and Viscosities of Binary Mixtures of n-Butanol with 2-Butanol, Isobutanol, and tert-Butanol from (303.15 to 343.15) K. J Chem Eng Data 55: 2310–2315. doi: 10.1021/je900722m

49. Łachwa J, Morgado P, Esperancüa José MSS, Guedes HJR, Lopes JNC, et al. (2006) Fluid-Phase Behavior of {1-Hexyl-3-methylimidazolium Bis(trifluoromethylsulfonyl) Imide, [C6mim][NTf2], +C2-C8 n-Alcohol} Mixtures: Liquid-Liquid Equilibrium and Excess Volumes. J Chem Eng Data 51: 2215–2221. doi: 10.1021/je060307z

50. Domınguez M, Gascon I, Valen A, Royo FM, Urieta JS (2000) Densities of (2-Butanol+n-Hexane+1-Butylamine)

at T = 298:15 And T = 313:15 K: Excess and Partial Excess Molar Volumes and Application of the Eras Model. J Chem Thermodyn 32: 1551–1568. doi: 10.1006/jcht.2000.0698

51. Huq F, Yu JQ (2002) Molecular Modeling Analysis: "Why is 2-Hydroxypyridine Soluble in Water but not 3-Hydroxypyridine?". J Mol Model 8: 81–86. doi: 10.1007/s00894-002-0073-1

52. Jorgensen WL, Chandrasekhas J, Madura JD, Impey RW, Klein ML (1983) Comparison of Simple Potential Functions for Simulating Liquid Water. J Chem Phys 79: 926–935. doi: 10.1063/1.445869

53. Dewar MJS, Storch DM (1985) Development and Use of Quantum Molecular Models. 75. Comparative Tests of Theoretical Procedures for Studying Chemical Reactions. J Am Chem Soc 107: 3898–3902. doi: 10.1021/ja00299a023

54. Pop E, Brewster ME (1997) Dimerization of Dexanabinol by Hydrogen Bonding Accounts for Its Hydrophobic Character. Intern J Quantum Chem 65: 1057–1064. doi: 10.1002/(sici)1097-461x(1997)65:6<1057::aid-qua4>3.0.co;2-u

55. Wilkes JS (2004) Properties of Ionic Liquid Solvents for Catalysis. J Mol Catal A: Chem 214: 11–17. doi: 10.1016/j.molcata.2003.11.029

56. Marsh KN, Boxall JA, Lichtenthaler R (2004) Room Temperature Ionic Liquids and Their Mixtures—a Review. Fluid Phase Equilib 219: 93–98. doi: 10.1016/j.fluid.2004.02.003

57. Zhu A, Wang J, Liu R (2011) A Volumetric and Viscosity Study for the Binary Mixtures of 1-Hexyl-3-Methylimidazolium Tetrafluoroborate with some Molecular Solvents. J Chem Thermodyn 43: 796–799. doi: 10.1016/j.jct.2010.12.027

Numerical Simulation of the Deflagration-to-Detonation Transition in Inhomogeneous Mixtures

Florian Ettner, Klaus G. Vollmer, and
Thomas Sattelmayer

Lehrstuhl für Thermodynamik, Technische Universität München, 85748 Garching, Germany

ABSTRACT

In this study the hazardous potential of flammable hydrogen-air mixtures with vertical concentration gradients is investigated numerically. The computational model is based on the formulation of a reaction progress variable and accounts for both deflagrative flame propagation and autoignition. The model is able to simulate versus time (x versus t) correlation is obtained from the photodiode measurements. The flame velocity between two subsequent photodiodes is calculated by applying a first order derivative:

$$v\left(x = \frac{x_i + x_{i+1}}{2}\right) = \frac{x_{i+1} - x_i}{t_{i+1} - t_i}.$$

(19)

Here, t_i and t_{i+1} represent the time at which the flame passes the photodiodes located at x_i and x_{i+1}, respectively. The same procedure is applied for evaluating the flame velocity in the numerical simulations.

Within the channel defined vertical concentration gradients can be generated. The overall amount of hydrogen is controlled via the partial pressure method. First, the air-filled channel is partially evacuated. Then, hydrogen is injected through several nozzles located at the top wall. The injection velocity is constant due to a choked nozzle upstream of the point of injection. The injection time defines the amount of hydrogen injected. Subsequently there is a defined time interval (waiting time t_w) during which diffusion takes place. Due to the strong density difference between hydrogen and air, a defined vertical concentration gradient is achieved while horizontal concentration gradients remain negligible. Finally the mixture is ignited by a spark plug. For a more detailed description of the experimental setup, the procedure of hydrogen injection, and mixture generation it is referred to the publications of Vollmer et al. [29, 60].

In order to determine the hydrogen distribution before ignition for many different hydrogen/air ratios and different waiting times, numerical simulations of the injection process have been conducted. Exemplary results for local hydrogen mole fraction over channel height at a waiting time $t_w = 3$ s (the strongest gradient under investigation) are shown in Figure 5. For waiting times $t_w >$ 30 s the mixture can be considered as homogeneous. The results of the injection simulations are stored as polynomials (hydrogen content versus channel height) which are used as initial conditions for the combustion simulations.

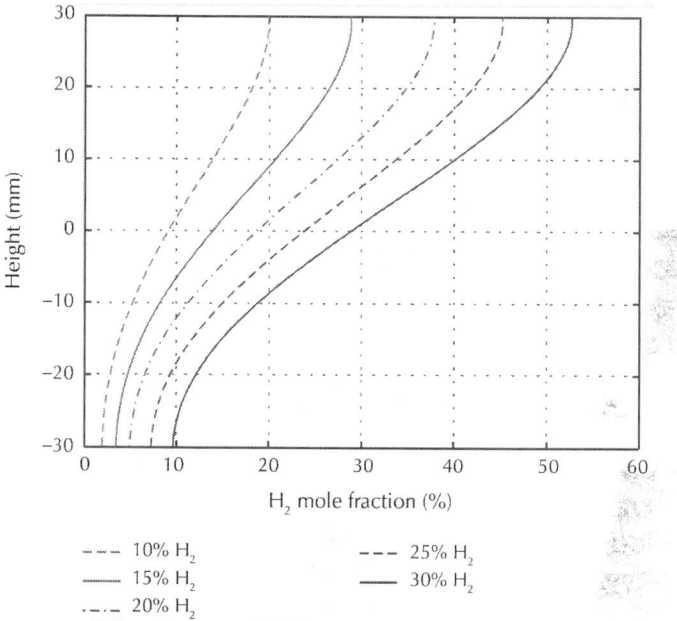

Figure 5: Local hydrogen content versus channel height after a waiting time of 3 s. The legend displays the overall hydrogen content of each mixture.

In the two-dimensional combustion simulations presented in the following section the channel is discretized with a uniform, rectangular grid of 2 mm grid spacing. Test runs showed that this resolution is the minimum resolution required to achieve grid independence with respect to the location of DDT. On coarser grids, DDT occurred mostly later or not at all. On finer grids, the location of DDT did not vary any more. However, at higher grid resolution, pressure peaks still got a little sharper. This should be kept in mind for the interpretation of the pressure plots shown in this paper.

Initially the fluid is at rest at a temperature of 293 K and a pressure of 1.01 bar. The boundary conditions are defined as adiabatic no-slip walls. Turbulence is modelled using the k-ω- SST model which is known for its good performance for both free-stream jets and wall-bounded flow [61, 62]. The initial hydrogen distribution either is homogeneous or corresponds to a concentration gradient of waiting time t_w = 3 s (see Figure 5).

Ignition is modelled by patching the site of ignition at $x=0$ with a burned mixture ($c=1$, see Figure 4). The initial turbulence is vanishingly small and consequently ξ equals unity so that s_T = s_L follows from (3). This means, the flame starts to propagate at laminar flame speed. However, turbulence is quickly generated by the flow itself so that the flame starts to accelerate. Test runs showed that the actual choice of initial turbulence variables is insignificant as long as $\xi=1$ is ensured. As the HLLC scheme gets unstable in the incompressible limit where no coupling between pressure and density exists, the first few time steps are calculated with a pressure-based solver [37] using the PISO scheme [40]. Before the flame reaches the first obstacle, the combustiondriven flow is usually strong enough to switch to the HLLC scheme that enables better shock capturing. Test runs showed that the transition between both schemes is smooth if it occurs while the maximum Mach number in the flow is in the range of 0.05 < Ma < 0.10.

RESULTS AND DISCUSSION

Experimental and numerical results for a homogeneous case with 15% hydrogen (volumetric) and blockage ratio BR = 30% are shown in Figure 6. It can be seen that the agreement between experiment and simulation is very good. The flame velocity rises continually in the obstructed part of the channel ($x \leq 2.05$ m). This can be attributed to the mutual amplification of combustion-induced expansion and turbulence generation due to interaction with obstacles. Shortly after passing the final obstacle the flame speed reaches a maximum and then decreases slowly. At $x \approx 4$ m the flame comes to a nearly complete rest before it accelerates again. This can be explained as follows: after passing the final obstacle, turbulence generation is diminished so that decelerating effects like friction outweigh the accelerating ones. The flame continuously gets slower. Simultaneously, while the flame has been consuming fresh gas, it generated pressure waves and displaced the unburned gas into the positive x direction. Shocks were generated that propagated towards the end wall from where they are being reflected. These reflected shocks now generate fluid flow in negative x direction. When the leading, backwards-running shock reaches the flame (this happens at $x \approx 4$ m), negative flow velocity and positive burning speed nearly cancel out so that the resulting net propagation velocity approaches zero. However, as there is still unburned gas in front of the flame, it recovers and accelerates again. The maximum flame propagation velocity of approximately 500 m/s indicates that no DDT occurred and the combustion process remained entirely deflagrative.

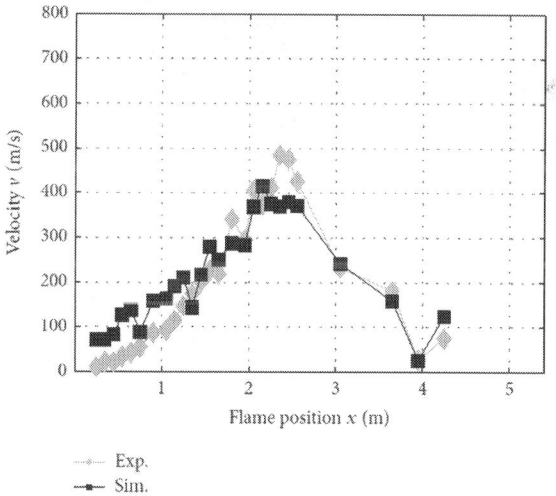

Figure 6: Flame propagation in a mixture with 15% H_2 (homogeneous).

Figure 7 shows the results for a mixture that contains an average hydrogen content of 15% as well, but with a vertical concentration gradient as shown in Figure 5. All other parameters are kept identical. In the early acceleration phase the flame velocity increases continually. This is in good agreement with the experiment. Then the flame is decelerated for the first time, due to a first shock front reflected from the end wall. The difference in the experiment can be attributed to the different ignition process: the spark generated by the spark plug in the experiment is considerably smaller than the ignition patch used in the simulation which is limited by the grid resolution. Thus the initial pressure wave generation in the experiment might be a little different from the initial pressure rise caused by the ignition in the numerical simulation. At the end of the obstacle region the flame speed peaks and then loses some driving force but eventually recovers. Although there is a considerable velocity difference between experiment and simulation in the unobstructed part, the final velocity is nearly the same. The pressures recorded by the sensor in the end wall reach extremely high values close to 120 bar (see Figure 8). In the homogeneous case, for comparison, the maximum pressure is in the range of 10 bar. The reason for the extreme pressure rise in the inhomogeneous case is revealed in Figure 9where the temperature and pressure distribution in the rear part of the explosion channel (4.9 m < x < 5.4 m) is displayed.

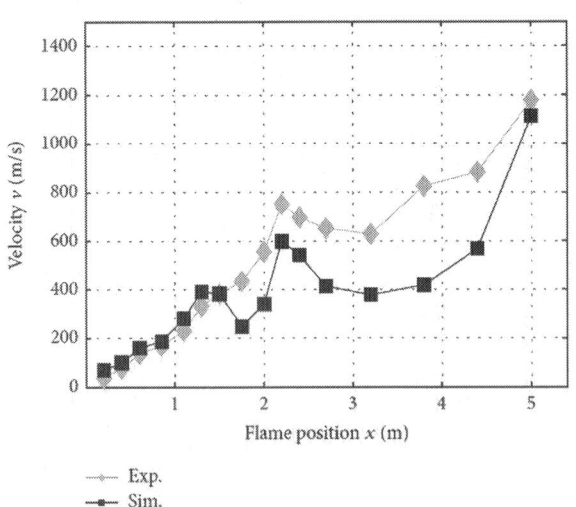

Figure 7: Flame propagation in a mixture with 15% H_2 (max. concentration gradient).

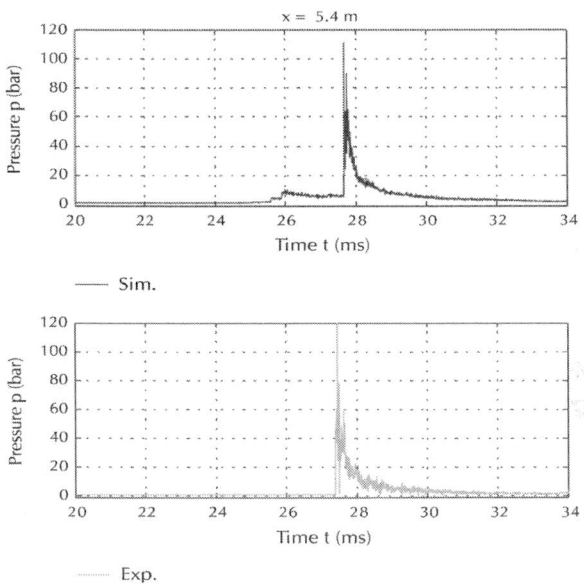

Figure 8: Pressure records from the sensor mounted on the end wall. Mixture with 15% H_2 (max. concentration gradient).

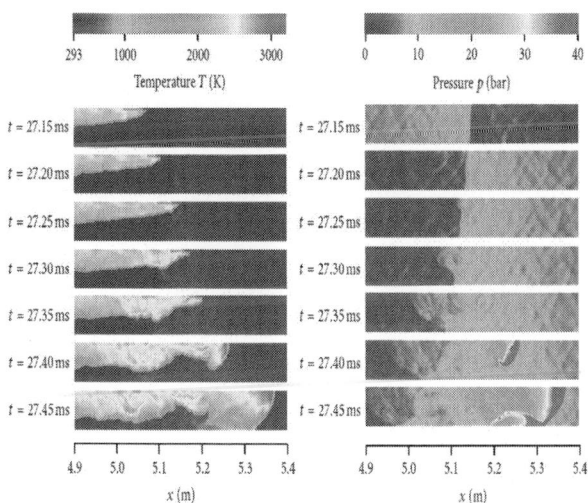

Figure 9: Visualization of a DDT caused by interaction of the flame with a reflected shock.

At t = 27.15 ms the flame approaches the end wall. Due to the inhomogeneous fuel distribution the flame is highly asymmetric and propagates mainly in the upper part of the channel. A leading shock has already been reflected from the end wall and moves towards the propagating flame. At t = 27.25 ms it reaches the flame. From this point onwards the flame burns into a precompressed mixture where the heat release rate is increased due to the increased density and increased laminar burning velocity (see (2) and (7)). The increased reaction rate leads to a strong pressure rise and causes an explosion at t = 27.40 ms. A radial detonation wave emanates from the explosion center and ignites the gas over the whole channel height. The newly formed detonation front runs towards the end wall where it causes an enormous pressure rise. This DDT mechanism has been suggested as one possible explanation for the high pressure loads observed in the experimental work of Eder [63]. In Eder's work, high pressure loads on the end wall of an explosion channel have been observed, but the flame velocity measurements indicated only a fast deflagration, not a detonation. As in the present simulation, the DDT in Eder's experiments obviously occurred so late (behind the final photo diode) that the DDT was not identified as one; only the high pressures on the end wall gave rise to speculation. Recent experimental investigations of Boeck et al. [64] support the conclusion that a DDT mechanism as identified in Figure 9 is responsible for the high pressure peak.

Due to the limited spatial resolution the present simulation does not resolve the interaction with the boundary layer. Moreover, it does not capture the shock-flame interaction in such a detailed manner as previous numerical studies on highly resolved grids (e.g., [12, 13]). Nevertheless the model is able to correctly predict the consequence of the backwards-running shock hitting the flame: an increased reaction rate due to precompression and intensified mixing which consequently triggers DDT.

From the pressure records in Figure 8 it can be concluded that there is a slight difference between experiment and simulation: the initial pressure rise in the simulation at $t \approx 26$ ms (caused by the reflection of the leading shock) does not appear in the experimental record. Due to the highly nonlinear dependence of ignition delay time on temperature and pressure, the higher propagation velocity in the experiment (Figure 7) is obviously sufficient to cause a strong autoignition quasi-instantaneously when the leading shock reaches the end wall. The resulting pressure load on the end wall, however, is nearly the same in experiment and simulation.

Increasing the hydrogen content leads to an earlier occurrence of DDT. At a hydrogen content of 25% (again with a concentration gradient as described in Figure 5) it can be seen from Figure 10 that the flame velocity rises continually to approximately 1000 m/s in the obstructed part of the channel and then suddenly jumps to 2500 m/s and finally relaxes to approximately 2000 m/s.

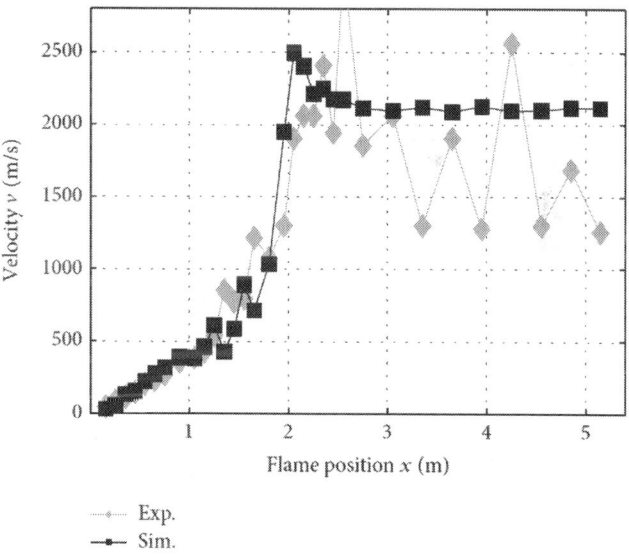

Figure 10: Flame velocity versus channel length for a mixture with 25% H_2 (max. concentration gradient).

This is a clear indication for the occurrence of a DDT with an initially overdriven detonation decaying to a Chapman-Jouguet detonation. The large fluctuations in the experimental velocity after the onset of DDT can be explained by small measurement errors in flame arrival time having a relatively large effect when the derivative

(19) is applied to the data. Using only the *x-t* diagram (Figure 11) as it is common in most publications does not reveal this difference.

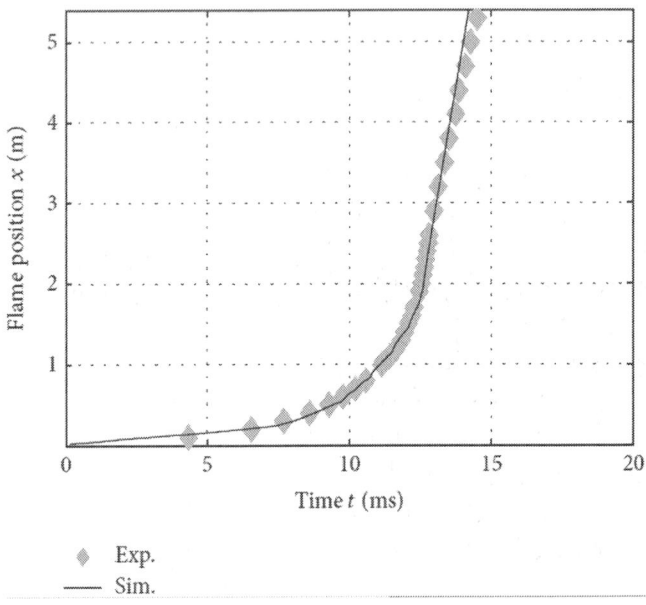

Figure 11: Flame position versus time for a mixture with 25% H$_2$ (max. concentration gradient).

The DDT process occurring in this case is visualized in Figure 12. At *t* = 12.44 ms, the flame approaches the final obstacle. The curved shock in front of the flame is reflected from the bottom wall by forming a Mach stem. At *t* = 12.45 ms, autoignition occurs behind the Mach stem. At *t* = 12.47 ms, the oblique shock hits the upper obstacle which initiates a second autoignition event. From there a circular detonation emanates and unites with the autoignition front from the lower part of the channel. While the detonation front moves through the gap between the obstacles into the unburned gas (*t* > 12.48 ms), the opposite front of the reaction wave ("retonation wave" [65]) runs backwards and consumes the remaining fresh gas in the lower part of the channel. It is important to note that the two autoignition kernels in Figure 12 are both well ahead of the flame but occur due to different reasons: the one on the bottom wall is due to shock compression ahead of the flame while the one on the upper wall occurs only due to reflection of the shock from the upper obstacle.

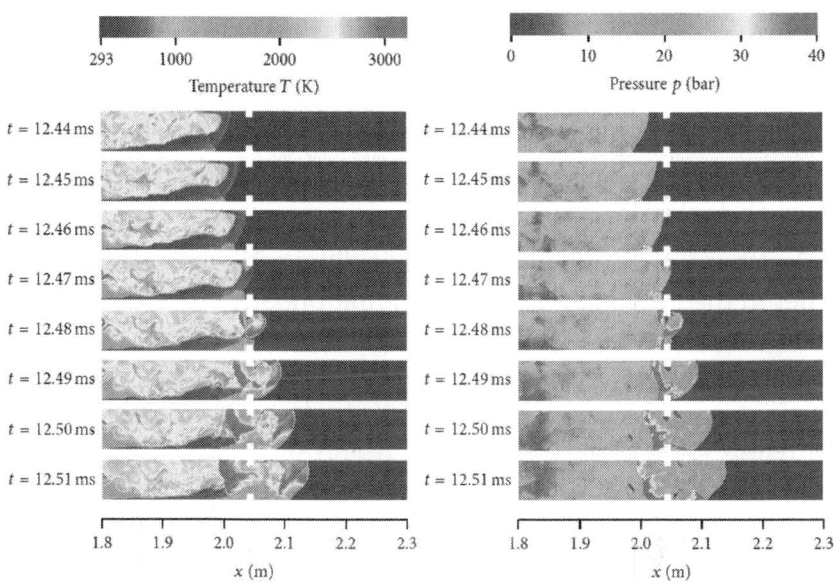

Figure 12: Visualization of a DDT in the vicinity of the final obstacle.

Another simulation with only six obstacles showed that the final obstacle was not necessary to achieve DDT. Instead, the autoignition occurring behind the Mach stem at $t = 12.45$ ms is sufficient to trigger DDT and is only amplified by the second autoignition event occurring on the upper obstacle. At lower fuel content (20% H_2) however, the seventh obstacle is required to obtain a DDT.

It is interesting to note that the first autoignition in Figure 12 occurs at the bottom wall where the mixture is leanest. This phenomenon can be explained by taking a closer look at the shock propagation: the leading shock approaches the final obstacle at a constant speed of $V \approx 1450$ m/s. Near the bottom wall the hydrogen content is 7% (see Figure 5) which results in a local speed of sound of $a = 356$ m/s. Thus, in the near vicinity of the bottom wall, the Mach stem can be seen as a normal shock propagating at Mach

increase or decrease the tendency towards DDT with the decisive factor being the obstacle geometry. If the obstacles are too large, they can lessen the DDT tendency as the majority of the strong pressure waves causing DDT are blocked. This is especially valid for an inhomogeneous mixture as shown in Figure 13, where the flame mainly burns in the upper part of the channel. In this case the obstacles are more obstructive than in a case with homogeneous mixture where the flame can be expected to propagate through the center of the channel where no obstacles are present.

Another question that has been addressed with the newly developed solver concerns the pressure loads that are caused by a steadily propagating detonation front, that is, after the occurrence of DDT. Therefore a look is taken at a detonation propagating in an unobstructed channel. First, this is demonstrated for a homogeneous mixture with 25% hydrogen. Pressure records are taken from the top and the bottom wall of the channel while a detonation passes. Test runs showed that the axial location of the pressure sensors did not influence the result any more as soon as a steadily propagating detonation was achieved. The result is shown in Figure 14. As the detonation front is nearly planar, the pressure records from the bottom and the top wall are virtually simultaneous. Upon arrival of the detonation front the pressure jumps to approximately 18 bar. The following expansion lets the pressure decrease slowly.

A completely different picture is found for a case with the same average hydrogen content, but a strong concentration gradient (Figure 15). As the leading shock is curved, it reaches the pressure sensors on the top wall earlier. They show maximum values of approximately 15 bar. On the bottom wall, however, a pressure of nearly 38 bar is reached. This is especially striking as the hydrogen content on the lower wall is only 7% and a homogeneous mixture with 7% hydrogen is basically nondetonable.

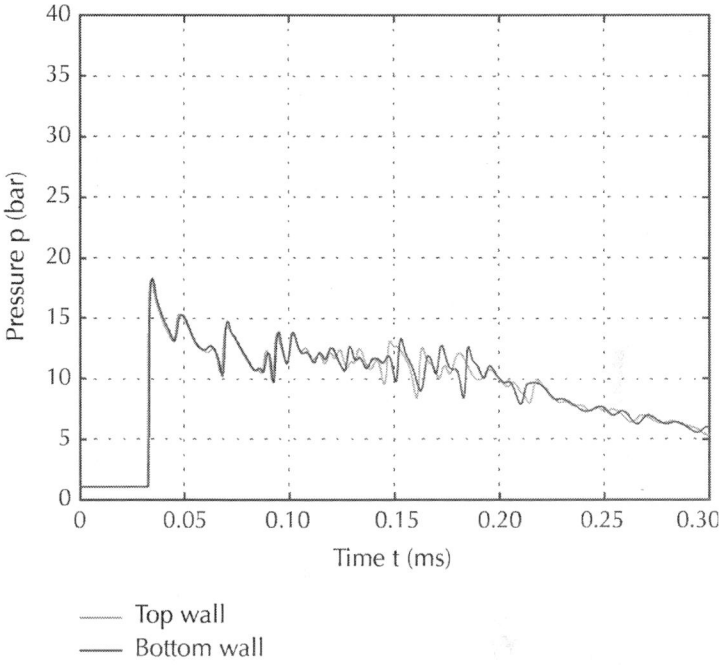

Figure 14: Pressure records from a steadily propagating detonation in a mixture with 25% H_2 (homogeneous).

Here, however, the lack of fuel does not lead to lower, but to higher, pressure loads. Again, the reason for this seeming paradox can be found in the particular structure of the leading shock front: on the bottom wall it is reflected via a Mach stem. Due to the lower speed of sound this causes a higher pressure rise on the bottom wall than on the top wall. After a short decline of the pressure, a second pressure rise is observed on both walls. This is due to secondary reflections of the leading shock that can be seen in the pressure field in Figure 16. Behind the secondary reflections the pressure equalizes so far that it drops simultaneously on the bottom and the top wall.

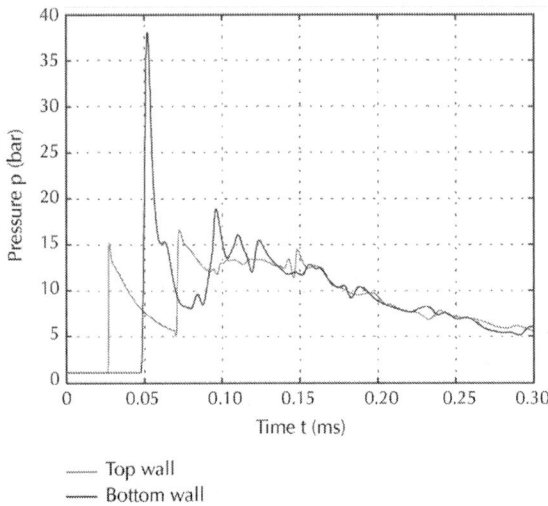

Figure 15: Pressure records from a steadily propagating detonation in a mixture with 25% H$_2$ (max. concentration gradient).

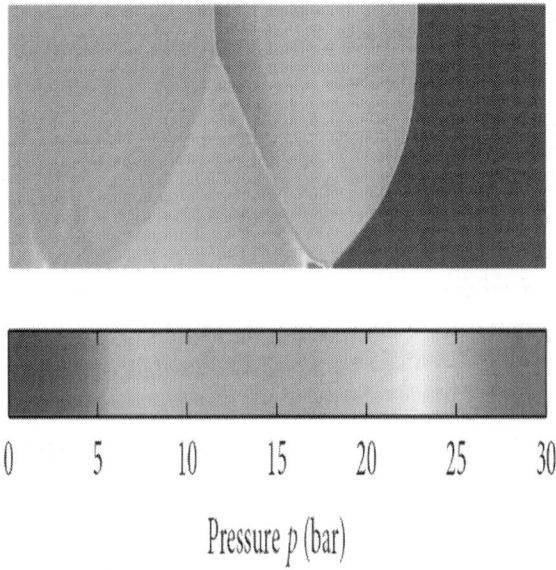

Figure 16: Pressure distribution in a steadily propagating detonation in a mixture with 25% H$_2$ (max. concentration gradient).

The results demonstrate that the pressure loads caused by a detonation in an inhomogeneous mixture can be considerably higher than in a homogeneous mixture of the same hydrogen content. Moreover, the location of the highest impact can be in fuel-lean regions. Further calculations showed that even if the hydrogen content on the bottom wall is reduced to zero, the maximum pressures observed there can still exceed those of the homogeneous mixture: the concentration gradient only needs to be strong enough to form a Mach stem. A simple method for predicting whether a detonation front in an inhomogeneous mixture develops a Mach stem can be found in [66].

SUMMARY, CRITICAL ANALYSIS, AND OUTLOOK

Motivated by the current lack of suitable tools for DDT-related safety studies [32], this paper presented a newly developed solver able to simulate flame acceleration, deflagration-to-detonation transition, and detonation propagation within a single run. The target was not to obtain detailed insight and maximum accuracy of the complex interaction between flow and reaction on microscopic scale, but to obtain a tool for engineering purposes that works on comparatively coarse grids and enables numerical safety studies at acceptable computational costs. The applicability to coarse grids is achieved by the inclusion of subgrid models. The agreement with experimental results is very good and the simulation gives additional insight into phenomena which cannot be easily observed in experiments. Although the simulations presented do not resolve all details of the flow, they are able to capture fundamental phenomena known from highly resolved simulations and experiments (e.g., [12, 13, 36]): DDT due to shock compression/Mach stem formation ahead of the flame, DDT due to shock reflection from obstacles, and DDT due to shock-flame interaction.

It has been found that concentration gradients, which are likely to occur in accident scenarios, can have a considerable effect on the nature of flame propagation. Depending on the enclosing geometry, the presence of a concentration gradient can decrease or increase flame propagation velocities, the probability of DDT, and the pressure loads associated with it. Thus, existing safety criteria developed for homogeneous mixtures can be inaccurate and nonconservative. Neither does a homogeneous mixture pose the highest threat regarding the probability of DDT nor does it cause the highest pressure loads. Due to gas dynamic phenomena within inhomogeneous mixtures, fuel-lean regions can be more DDT-prone than stoichiometric or rich regions. This should be taken into account in future safety studies.

However, although the general agreement with experiments is good, it has to be kept in mind that all results have been gained on relatively coarse grids, without resolving the induction distance between shock and reaction in a detonation front. Boundary layers are not resolved either and they are known to have an effect on the onset of detonations. Moreover, all the results presented in this study have been obtained on 2D grids. Therefore, the authors have deliberately chosen a geometry which is relatively wide (300 mm) compared to its height (60 mm). Nevertheless transversal waves are expected to play a role in flame acceleration and the onset of DDT. While this project has already finished, a follow-up project has started where 3D simulations are conducted [67] and also simulations in large, complex geometries with the aim of reproducing realistic accident scenarios. Approaches are being developed to use even coarser grids by applying subgrid models not only to shock propagation as shown in this paper, but also to deflagrative flame propagation.

The advantage of the solver developed and its implementation in OpenFOAM is that it is not limited to structured grids and thus can be applied to intricate geometries using unstructured grids as well. The solver and its source code are made freely available to the public [68].

ACKNOWLEDGMENTS

This project is funded by the German Federal Ministry of Economics and Technology on the basis of a decision of the German Bundestag (Project no. 1501338) which is gratefully acknowledged. The authors would like to thank Oliver Borm for sharing a Riemann solver code within the framework of OpenFOAM which provided a valuable basis for this study.

REFERENCES

1.   Y.-L. Liu, J.-Y. Zheng, P. Xu et al., "Numerical simulation on the diffusion of hydrogen due to high pressured storage tanks failure," Journal of Loss Prevention in the Process Industries, vol. 22, no. 3, pp. 265–270, 2009.

2.   B. P. Xu, J. X. Wen, S. Dembele, V. H. Y. Tam, and S. J. Hawksworth, "The effect of pressure boundary rupture rate on spontaneous ignition of pressurized hydrogen release," Journal of Loss Prevention in the Process Industries, vol. 22, no. 3, pp. 279–287, 2009.

3.   W. Breitung and P. Royl, "Procedure and tools for deterministic analysis and control of hydrogen behavior in severe accidents," Nuclear Engineering and Design, vol. 202, no. 2-3, pp. 249–268, 2000. ·

4.   M. Manninen, A. Silde, I. Lindholm, R. Huhtanen, and H. Sjövall, "Simulation of hydrogen deflagration and detonation in a BWR reactor building," Nuclear Engineering and Design, vol. 211, no. 1, pp. 27–50, 2002.

5.   H. Dimmelmeier, J. Eyink, and M.-A. Movahed, "Computational validation of the EPR combustible gas control system," Nuclear Engineering and Design, vol. 249, pp. 118–124, 2012.

6.   A. G. Venetsanos, D. Baraldi, P. Adams, P. S. Heggem, and H. Wilkening, "CFD modelling of hydrogen release, dispersion

and combustion for automotive scenarios," Journal of Loss Prevention in the Process Industries, vol. 21, no. 2, pp. 162–184, 2008.

7. W. G. Houf, G. H. Evans, E. Merilo, M. Groethe, and S. C. James, "Releases from hydrogen fuel-cell vehicles in tunnels," International Journal of Hydrogen Energy, vol. 37, no. 1, pp. 715–719, 2012.

8. S. B. Dorofeev, "Flame acceleration and explosion safety applications," Proceedings of the Combustion Institute, vol. 33, no. 2, pp. 2161–2175, 2011.

9. S. B. Margolis and B. J. Matkowsky, "Nonlinear stability and bifurcation in the transition from laminar to turbulent flame propagation," Combustion Science and Technology, vol. 34, no. 1–6, pp. 45–77, 1983. ·

10. G. I. Sivashinsky, "Instabilities, pattern formation and turbulence in flames," Annual Review of Fluid Mechanics, vol. 15, pp. 179–199, 1983.

11. E. S. Oran and A. M. Khokhlov, "Deflagrations, hot spots, and the transition to detonation,"Philosophical Transactions of the Royal Society A: Mathematical, Physical and Engineering Sciences, vol. 358, no. 1764, pp. 3539–3551, 2000.

12. E. S. Oran and V. N. Gamezo, "Origins of the deflagration-to-detonation transition in gas-phase combustion," Combustion and Flame, vol. 148, no. 1-2, pp. 4–47, 2007.

13. V. N. Gamezo, T. Ogawa, and E. S. Oran, "Flame acceleration and DDT in channels with obstacles: effect of obstacle spacing," Combustion and Flame, vol. 155, no. 1-2, pp. 302–315, 2008.

14. M. A. Liberman, A. D. Kiverin, and M. F. Ivanov, "On detonation initiation by a temperature gradient for a detailed chemical reaction models," Physics Letters A: General, Atomic and Solid State Physics, vol. 375, no. 17, pp. 1803–1808, 2011.

15. M. C. Gwak and J. J. Yoh, "Effect of multi-bend geometry on deflagration to detonation transition of a hydrocarbon-air

mixture in tubes," International Journal of Hydrogen Energy, vol. 38, no. 26, pp. 11446–11457, 2013.

16.   P. Hwang, R. P. Fedkiw, B. Merriman, T. D. Aslam, A. R. Karagozian, and S. J. Osher, "Numerical resolution of pulsating detonation waves," Combustion Theory and Modelling, vol. 4, no. 3, pp. 217–240, 2000.

17.   G. J. Sharpe, "Transverse waves in numerical simulations of cellular detonations," Journal of Fluid Mechanics, vol. 447, pp. 31–51, 2001.

18.   G. Cael, H. D. Ng, K. R. Bates, N. Nikiforakis, and M. Short, "Numerical simulation of detonation structures using a thermodynamically consistent and fully conservative reactive flow model for multi-component computations," Proceedings of the Royal Society A: Mathematical, Physical and Engineering Sciences, vol. 465, no. 2107, pp. 2135–2153, 2009.

19.   K. Mazaheri, Y. Mahmoudi, and M. I. Radulescu, "Diffusion and hydrodynamic instabilities in gaseous detonations," Combustion and Flame, vol. 159, no. 6, pp. 2138–2154, 2012.

20.   J. M. Powers and S. Paolucci, "Accurate spatial resolution estimates for reactive supersonic flow with detailed chemistry," AIAA Journal, vol. 43, no. 5, pp. 1088–1099, 2005.

21.   M. Manninen, R. Huhtanen, I. Lindholm, and H. Sjövall, "Hydrogen in BWR reactor building," inProceedings of the 8th International Conference on Nuclear Engineering (ICONE ‹00), Baltimore, Md, USA, 2000.

22.   D. M. Prabhudharwadkar, K. N. Iyer, N. Mohan, S. S. Bajaj, and S. G. Markandeya, "Simulation of hydrogen distribution in an Indian Nuclear Reactor Containment," Nuclear Engineering and Design, vol. 241, no. 3, pp. 832–842, 2011.

23.   S. B. Dorofeev, A. S. Kochurko, A. A. Efimenko, and B. B. Chaivanov, "Evaluation of the hydrogen explosion hazard," Nuclear Engineering and Design, vol. 148, no. 2-3, pp. 305–316, 1994.

44. A. Favre, "Equations des gaz turbulents compressibles," Journal de Mécanique, vol. 4, pp. 361–390.

45. C. Chen, J. J. Riley, and P. A. McMurtry, "A study of Favre averaging in turbulent flows with chemical reaction," Combustion and Flame, vol. 87, no. 3-4, pp. 257–277, 1991.

46. M. Brandt, W. Polifke, B. Ivancic, P. Flohr, and B. Paikert, "Auto-ignition in a gas turbine burner at elevated temperature," in Proceedings of the ASME Turbo Expo, pp. 195–205, Atlanta, Ga, USA,, June 2003.

47. O. Colin, A. Pires da Cruz, and S. Jay, "Detailed chemistry-based auto-ignition model including low temperature phenomena applied to 3-D engine calculations," Proceedings of the Combustion Institute, vol. 30, no. 2, pp. 2649–2656, 2005.

48. J.-B. Michel, O. Colin, and C. Angelberger, "On the formulation of species reaction rates in the context of multi-species CFD codes using complex chemistry tabulation techniques," Combustion and Flame, vol. 157, no. 4, pp. 701–714, 2010.

49. H. G. Weller, G. Tabor, A. D. Gosman, and C. Fureby, "Application of a flame-wrinkling LES combustion model to a turbulent mixing layer," Symposium (International) on Combustion, vol. 27, pp. 899–907, 1998.

50. K. N. C. Bray, "Complex chemical reaction systems," in Chapter Methods of Including Realistic Chemical Reaction Mechanisms in Turbulent Combustion Models, vol. 47 of Springer Series in Chemical Physics, pp. 356–375, Springer, 1986.

51. V. L. Zimont and A. N. Lipatnikov, "A numerical model of premixed turbulent combustion of premixed gases," Chemical Physics Reports, vol. 14, pp. 993–1025, 1995.

52. F. Ettner, Effiziente numerische simulation des deflagrations-detonations-Übergangs [Ph.D. thesis], TU München, 2013.

53. A. A. Konnov, "Remaining uncertainties in the kinetic mechanism of hydrogen combustion," Combustion and Flame, vol. 152, no. 4, pp. 507–528, 2008. ·

54. W. Polifke, P. Flohr, and M. Brandt, "Modeling of inhomogeneously premixed combustion with an extended TFC model," Journal of Engineering for Gas Turbines and Power, vol. 124, no. 1, pp. 58–65, 2002.

55. S. R. Turns, An Introduction to Combustion, McGraw-Hill, 2000.

56. D. Goodwin, "Cantera: an object-oriented software toolkit for chemical kinetics, thermodynamics and transport processes," 2009, http://code.google.com/p/cantera.

57. M. Ó Conaire, H. J. Curran, J. M. Simmie, W. J. Pitz, and C. K. Westbrook, "A comprehensive modeling study of hydrogen oxidation," International Journal of Chemical Kinetics, vol. 36, no. 11, pp. 603–622, 2004.

58. L. Tosatto and L. Vigevano, "Numerical solution of under-resolved detonations," Journal of Computational Physics, vol. 227, no. 4, pp. 2317–2343, 2008.·

59. J. D. Anderson, Modern Compressible Flow, McGraw-Hill, 2004.

60. K. G. Vollmer, F. Ettner, and T. Sattelmayer, "Influence of concentration gradients on flame acceleration in tubes," in Proceedings of the 8th International Symposium on Hazards, Prevention and Mitigation of Industrial Explosions, Yokohama, Japan, 2010.

61. F. R. Menter, "Two-equation eddy-viscosity turbulence models for engineering applications," AIAA Journal, vol. 32, no. 8, pp. 1598–1605, 1994.

62. F. R. Menter, "Review of the shear-stress transport turbulence model experience from an industrial perspective," International Journal of Computational Fluid Dynamics, vol. 23, no. 4, pp. 305–316, 2009. ·

63. A. Eder, Brennverhalten schallnaher und überschall-schneller Wasserstoff-Luft Flammen [Ph.D. thesis], TU München, 2001.

64. L. R. Boeck, J. Hasslberger, F. Ettner, and T. Sattelmayer, "Investigation of peak pressures during explosive combustion of inhomogeneous hydrogen-air mixtures," in Proceedings

of the 7th International Fire and Explosion Hazards Seminar, Providence, RI, USA, 2013.

65. A. K. Oppenheim, A. J. Laderman, and P. A. Urtiew, "The onset of retonation," Combustion and Flame, vol. 6, pp. 193–197, 1962.

66. F. Ettner, K. G. Vollmer, and T. Sattelmayer, "Mach reflection in detonations propagating through a gas with a concentration gradient," Shock Waves, vol. 23, pp. 201–206, 2013.

67. J. Hasslberger, F. Ettner, L. R. Boeck, and T. Sattelmayer, "2D and 3D flame surface analysis of flame acceleration and deflagration-to-detonation transition in hydrogenair mixtures with concentration gradients," in Proceedings of the 24th International Conference on the Dynamics of Explosions and Reactive Systems (ICDERS ‹13), Taipei, Taiwan, 2013.

68. F. Ettner and T. Sattelmayer, ddtFoam, 2013, http://sourceforge.net/projects/ddtfoam.

7

New Simple Indices for Risk Assessment and Hazards Reduction at the Conceptual Design Stage of a Chemical Process

Mohammad Hossein Ordouei[a], Ali Elkamel[a], and Ghanima Al-Sharrah[b]

[a]Department of Chemical Engineering, University of Waterloo, 200 University Avenue West Waterloo, Ontario, Canada N2L 3G1

[b]Department of Chemical Engineering, Kuwait University, Safat, Kuwait

ABSTRACT

Inherent safety has been of great interest to regulators, process designers and investors. The idea behind this is that a process design is more economic when it is inherently safer. Inherent safety is known as the safety intrinsic to a process; the spirit of which is to mitigate hazards within the process. It is also possible to achieve inherently safer design by diminishing the hazards in multi-component streams during process design. Hazards reduction during the design phase is a challenging task. A decrease in hazards in a process design not only improves process safety, but also protects the environment from potential impacts of the process. Current methodologies for risk assessment at the conceptual design stage of a chemical process need detailed process data, which is usually unavailable at such a phase. This paper presents simple new indices that require minimum data for risk evaluation of chemical processes at the conceptual design phase. The indices are applied to a hydrogenation case study to choose inherently safer designs among different alternatives. As an important result, total capacity of a process among other design array does not suffice for decision making unless the mass fraction of hazards in product streams are appreciably low.

INTRODUCTION

An accident in a chemical manufacturing plant is not only harmful to the plant; it can also be an irreparable spoil for the reputation of the licensing company who has designed the chemical process. This fact reveals that it is imperative to alleviate possible risks to process safety during the design phase. The tie between process design and the risk to process safety is not new; any kinds of design modifications and/or the development of operating instructions result in risk reduction within the process plant; e.g. purification of raw material, centralization of hazardous chemicals in safe containers or bags and transformation of the hazardous chemicals to benign materials (Carson and Mumford, 2002). There are several qualitative and quantitative methods to estimate the risks associated with a chemical process; however, few of them can be used in conceptual design.

Chemicals, in general, are the main source of fire, explosion, toxicity and corrosion hazards. About two third of impacts were initiated mainly by explosion compared to fire (Lees, 1996); however, toxicity is more influential on the number of affected people compared to fire and explosion (Belke, 2000). Thus, it is vital to pay close attention to the chemical toxicity for the risk assessment during the primitive step of process design, especially in the absence of detailed information about the process.

Hazard is an intrinsic chemical or physical property of a material or a system or a process, which can be detrimental to human, plant, equipment, and environment. Hazard and risk have two distinctive concepts (Canadian Centre for Occupational Health and Safety, 2009). A "hazard" refers to the potential of negative consequences on personnel's health or company's equipment and property, while a "risk" is defined as the probability of the hazard, which results in adverse effects on the human, the property or the equipment. Hence, the risk is generally a function of two factors; frequency and consequences:

$$\text{Risk Assessment} = f(\text{Frequency, Consequences}) \tag{1}$$

This relationship has persuaded Marhavilas et al. (2011) to develop a model called decision matrix risk assessment (DMRA). It is also widely being used by other researchers and engineers (Reniers et al., 2005, Woodruff, 2005, Henselwood and Phillips, 2006 and Marhavilas and Koulouriotis, 2008).

Researchers have made several attempts to provide simple methodologies for the evaluation of potential risk to a process safety. A simple risk index is a mathematical model to be employed in the primitive stage of planning in chemical plants, easily applicable in process plants, include industrial experience and require general plant (Al-Sharrah et al., 2007).

The review of such methodologies is out of the scope of this paper; examples include (but not limited to) STEP, HAZOP, What-If Analysis, PRA, Checklist Analysis, SA, TA, FTA, DMRA,

inherent safety in a chemical process; i.e. hazards alleviation instead of employing protective devices (Heikkilä, 1999). Consequently, the question of the severity and the likelihood of an accident can be addressed when hazards have been identified.

Inherently safer design (ISD) is highly supported by the availability of simple indices that can be used at early stages of design. ISD is an approach to address the risks of hazardous chemicals to human, environment and process plant during design and manufacturing phases of a process (Hendershot, 2011a). The term ISD was first introduced in the 1970s after the big disaster in Flixborough, UK, in 1974; however, the concept of inherent safer design (ISD) is not new. It has been used since Stone Age when cave inhabitants decided to move up to a higher level of the cave to diminish the risk of flood, while they could reduce the risk by either of dike (engineering control) and monitoring the level of river (administrative control).

Today, more researchers and engineers are becoming familiar with ISD through new publications and training such as the relevant course provided by AIChE:

https://www.aiche.org/ccps/resources/education/courses/ch800/inherently-safer-design

Together with engineering and administrative controls, ISD is able to manage the risks of a process efficiently. There are four strategies to design an inherently safer process (CCCP, 2009):

- Substitution of hazardous chemicals with benign materials.
- Minimization of hazardous materials.
- Moderating the process by dilution, refrigeration etc.
- Simplification of operation by reducing the potential errors such as using interlocking commands for process control equipment.

It is now possible to replace toxic chemicals in off-shore oil and gas facilities during conceptual design in order to design an inherently safer process at optimum cost and minimum acceptable risk (Khan and Amyotte, 2002). The concept of inherent risk assessment has been used for the integration of risk quantification into HYSYS process simulator (Leong and Shariff, 2008 and Shariff and Leong, 2009).Cordella et al. (2009) have provided a comprehensive method for screening the design of inherently safer processes based on categorization of hazards with respect to human, ecosystem, and environmental media contamination.

A process design is performed in three steps: basic (e.g. conceptual) design, front end engineering, and detailed design. In each step, several technical documents are generated by corresponding departments, but risk assessment is usually accomplished at the final step of process design.

In effect, conceptual design plays a decisive role in minimizing the risks of a process since all other design steps are based on this phase. It turns out that the impact of decisions is extremely high at the conceptual design stage (Lewin, 2004) but can be minimized if the process is inherently (internally) safe and well conceptualized since protective and control devices (external safety) would be either eliminated or have smaller sizes. In other words, a conceptual design encompasses less decision making impacts when it comes to inherently safer processes (Kletz, 2001) due to selection of cheaper materials for piping and equipment. Thus, inherent safety at low expenditure can be achieved during conceptual design.

The substitution of hazardous chemicals with benign materials is not the only way for inherently safer design. For instance, raw materials used in petrochemical and refinery plants predominantly contain flammable and toxic hydrocarbons and almost impossible to be replaced by other chemicals. Hence, the inherent safety in this case is accomplished by minimizing the mass fraction of hazards in product as well as waste streams (source reduction) resulting in "safety improvements" and "environmental protection".

Unfortunately, the risks of chemical hazards in design phase have been overlooked in almost all of conventional process design leading to generation of large amounts of waste as pollutant source (EPA, 2012), whilst the most convenient time for effective source reduction falls in process design phase (Tchobanoglous, 2009). Traditionally, the safety and environmental considerations were left to designer experience at the initial phase of the design (Koller et al., 1999).

The objective of this paper is (1) to develop simple and new indices for evaluation of the risks associated with hazards in process streams to be implemented at the conceptual design stage with minimum available data, (2) to apply the risk indices to hydrogenation process design alternatives, (3) and finally to analyze and discuss the results. These risk indices have direct relationship with the mass fraction of hazardous chemicals (x) in the process streams and help process and safety designers to "protect the environment" while "decreasing the process risks" by minimizing the hazards mass fraction (x).

The inherent safety is being used in practical situations by a number of researchers (Heikkilä et al., 1996,Khan and Abbasi, 1998, Khan and Amyotte, 2002, Shariff et al., 2006, Leong and Shariff, 2008, Cordella et al., 2009, Shariff and Leong, 2009, Hendershot, 2011a and Hendershot, 2011b).

A NEW RISK INDEX FOR USE IN CONCEPTUAL DESIGN

From the preceding sections, we learned that the existing risk assessment methodologies are comprehensive, time consuming and requires detailed process data and therefore, they are not suitable for conceptual design. Hence, there is still a demand to new simple indices for use in the conceptual design phase for evaluation of different proposals when new or retrofitting processes are concerned. Supposing the situations like tendering a chemical process project where a number of proposals are sent to a client asserting to design the safest process; each proposal provides minimum process data such as general process description, simple block diagram, chemical compositions, mass fraction of hazardous materials in product and waste streams, such new indices shall only employ the above data in addition to chemical toxicity, process inventory, and the history of previous accidents in corresponding processes for risk estimation and the screening purpose of all proposals.

Although the recently improved risk index given by Eq. (2) is applicable in the majority of such situations, it gives unfavorable results in the following instances:

- The index adds up all streams within the entire process (instead of products and waste streams). Then, it multiplies the result by one month production of the process as maximum inventory, which results in an unrealistic increase in the risks associated with the process.

- In case of comparing two processes with the same flow rate of fresh feed; i.e. one with waste recycling and the other without, the number of streams within the recycling process will be more and consequently, the associated risks will vividly increase resulting in rejection of all types of recycling designs. This is an unacceptable result since the waste recycling, source reduction and prevention of waste generation are all remedies to minimize pollution and hazards within manufacturing plants (Pankratz, 2001).

- In Eq. (2), the term "*Size*" of a plant most probably equals to three as stated earlier. Although most reactants and products go through these three sections, tripling the estimated risks of the process may be misleading. For instance, in the reaction section the reactants transform to products in the course of reaction and will not exist in other sections and therefore, tripling the corresponding hazard leads to a misleading or even the wrong result. Furthermore, it is possible to isolate any risks in each section or equipment by control valves or block valves and so on. Hence, the term "*Size*" in such cases should equal to unity (one process).

- When two distinct processes with the same risks are concerned, Eq. (2) leads to no result unless both of them handle single component streams. For multi-component streams the purity of the product streams is so vital in choosing the inherently safer design. In the latter case, both severity and mass fractions of hazardous chemicals (impurities) are important.

The new risk indices proposed in this paper resolve the above deficiencies by evaluation of the risks associated with the mass flow rate of hazardous chemical components in both product and waste streams within a process instead of risk assessment associated with the whole process.

Based on the fundamental Eq. (1), the new indices are also function of either of accident frequency and hazard effects of chemical components.

A chemical process plant may have multiple product streams; therefore, the risk for all streams can be estimated as follows:

$$(R.I)^P = \sum_i \sum_j M_j \times f_i \times H_i \times x_{i,j}$$
(3)

Similarly, the risk for more than one waste streams can be calculated by following equation:

$$(R.I)^W = \sum_i \sum_j M_j \times f_i \times H_i \times x_{i,j}$$
(4)

where R.I is an abbreviation for Risk Index, superscripts P and W denote product and waste streams, respectively. So, $(R.I)^P$ and $(R.I)^W$ express the impacts of the calculated risks in "number of affected people per year", which is the maximum potential risks attributed to the total product streams and the total waste streams, respectively. Subscripts i and j designate the "chemicals within the streams" and the "streams within the process", respectively. M_j stands for the mass in tons chemical released to the environment and is defined as maximum one month production of the process plant or one month inventory (Couper et al., 2005), which can be calculated from design basis. f_i represents the frequency of accident for chemical component i in "number of accidents per year" (Belke, 2000).

H_i denotes the hazard effects of chemical i, in "number of people affected per ton of chemical released" to the environment (ARIP,

1999). $x_{i,j}$ is the mass fraction of component i in stream j (i, j=1, 2, ...). Table 1 presents the data of "H" and "f" for some chemicals.

Table 1: Data for the severity and the likelihood of accidents for some chemicals. Adapted from Al-Sharrah et al. (2007).

No.	Chemical	*H* people affected per ton chemical released	*f* a Number of accident per year
1	Acetaldehyde	0.1202	0.008
2	Acetic acid	0.0229	0.038
3	Acrolein	0.5763	0.064
4	Acrylic acid	0.0561	0.038
5	Acrylonitrile	0.4224	0.042
6	Ammonia	0.1357	0.016
7	Benzene	0.1465	0.008
8	Butadiene	0.1233	0.013
9	Carbon tetrachloride	0.1827	0.056
10	Chlorine	0.8105	0.022
11	Cumene	0.0742	0.008
12	Ethane	0.1526	0.014
13	Ethyl benzene	0.0451	0.008
14	Formaldehyde	1.8414	0.009
15	Hydrogen chloride	0.4273	0.06
16	Hydrogen cyanide	5.9972	0.064
17	Hydrogen fluoride	0.0116	0.064
18	Nitric acid	0.2298	0.038
19	Pentane	0.1515	0.013
20	Phenol	0.0002	0.008
21	Phosphoric acid	0.0133	0.038
22	Styrene	0.4484	0.008

23	Sulfuric acid	0.0149	0.038
24	Toluene	0.0747	0.008
25	Vinyl acetate	0.1866	0.042
26	Vinyl chloride	0.0337	0.042
27	Xylene	0.2348	0.008
28	CH_4	N/A	0.03
29	CH_2Cl_2	0.21	0.04
30	Chloroform	0.02	0.04
31	CCl_4	0.1827	0.056
32	Cl_2	0.8105	0.022
33	HCl	0.4273	0.06
34	CH_3Cl	0.07	0.04
35	Iso-butane	0.0832	0.013
36	n–Butane	0.3296	0.013
37	Iso-pentane	0.1761	0.013
38	n-Pentane	0.249042	0.013
39	Iso-butene	0	0.013
40	Iso-octene	0.3199	0.013
41	Iso-octane	0.3535	0.013
42	1-Butene	0.1996	0.013
43	Styrene	0.4484	0.008
44	Ethyl-benzene	0.0451	0.008
45	Benzene	0.1465	0.008
46	Toluene	0.0747	0.008
47	Hydrogen	0	0.013

[a]Considering the number of processes to be equal to unity (i.e. one process), frequency of accidents (Belke, 2000) can be modified as reported in column four in number of accidents per year.

$(R.I)^P$ is an acceptable risk since it is associated with the product streams, which corresponds to company's profits. However, the process design with the least $(R.I)^P$ is enviable when other design factors such as design cost favor that design. In case of the possibility of substitution of a hazardous material in a process with a benign substance at the same production rate, the $(R.I)^P$ will be less for the same process so; the process will be inherently safer.

Conventionally, chemical process designers considered a storage tank at the plant battery limit (B.L.) of maximum capacity of one month production (Couper et al., 2005) as inventory aiming to prevent downstream plants from shut-down in case of failing the upstream process, because the tank at B.L. would give enough time to operation or maintenance staffs for troubleshooting while supplying feed to downstream plants.

The authors' observations and experience show that recently, the inventory at B.L. has been superseded in most processes especially in refineries and petrochemical complexes. Therefore, in this paper one month production of product and waste streams is considered as maximum inventory.

In some of traditional risk assessment methodologies, the metrics such as population densities and the types of land were used as the basis for risk management in the construction step of an industry (Henselwood and Phillips, 2006). This approach has only been used in the construction phase and cannot be used in the conceptual design stage, since the population distribution becomes important when the plant location is concerned but neither plant location nor population density is the scope of conceptual design. Furthermore, in case of an accident in a process, the population living around the plant area is a key point to estimate the maximum number of affected people; however, the population density has no influence on total risk associated with a process safety itself, since the risk is a function of frequency and consequences (or likelihood and severity).

Above all, in some incidents leading to release of toxic materials, zero people has been affected since nobody was in the proximity of the accident when it happened (Al-Sharrah et al., 2007). This divulges that zero affected people means that the process involved neither is inherently safe nor is it influenced by population density.

The total risk, $(R.I)^T$, is the summation of the risks associated with products and wastes:

$$(R.I)^T = (R.I)^P + (R.I)^W \tag{5}$$

Dividing both sides of Eq. (5) by annual production capacity of the process plant in ton per year, results in a new risk index as follows:

$$(R.I)^T/j_{\sum P_j} = (R.I)^P/j_{\sum P_j} + (R.I)^W/j_{\sum P_j} \tag{6}$$

where subscript j denotes the stream number of products. The left hand side of Eq. (6) represents the total risks per ton products. The first term of the right hand side of Eq. (6) represents the risks associated with product streams per ton products and the second term represents the risks associated with waste streams per ton products.

The normalized index, Eq. (6), is independent from the process size and enables a process engineer to compare two processes with different production capacities. Eqs. (3), (4) and (6) are all of great magnitude in ranking of process designs from a safety point of view.

When other circumstances in two or more designs are the same, the term $(R.I)^P/J_{\sum P_j}$ would merely be eminent for the ranking of the process designs.

ILLUSTRATIVE CASE STUDIES

Case study 1: Chlorination of Methane

In this case study, the chlorination of methane was modeled by the Aspen HYSYS process simulator. The chlorination of methane is a heterogeneous catalytic reaction, which takes place in four steps:

$$CH_4 + \frac{1}{2}Cl_2 \rightarrow CH_3Cl + HCl \tag{7}$$

$$CH_3Cl + \frac{1}{2}Cl_2 \rightarrow CH_2Cl_2 + HCl$$

(8)

$$CH_2Cl_2 + \frac{1}{2}Cl_2 \rightarrow CHCl_3 + HCl$$

(9)

$$CHCl_3 + \frac{1}{2}Cl_2 \rightarrow CCl_4 + HCl$$

(10)

$$CH_4 + 2Cl_2 \rightarrow CCl_4 + 4HCl$$

(11)

The chlorination rate constant varies significantly; for instance, the chlorination of benzene (Levenspeil, 1999):

$$C_6H_6 + Cl_2 \rightarrow C_6H_5Cl + HCl \quad k = 0.412$$

(12)

And the chlorination of C—H bond of cyclopentane at 40°C in gas phase (Denisov, 1974):

$$C_5H_{10} + Cl_2 \rightarrow C_5H_9Cl + HCl \quad \log k = 10.08$$

(13)

The reactions (7)–(10) are complex series-parallel reactions, which means series with respect to chlorinated species and parallel with respect to Cl_2 (Missen et al., 1999).

The factors influencing the kinetics of the methane chlorination include:

- Gas flow regime plus gas-phase thermal reaction (Rozanov and Treger, 2010),
- Catalyst and CH_4/Cl_2 ratio effects (Bucsi and Olah, 1992),
- Inert diluents effects (e.g. N_2) on the photo-chlorination of methane and its selectivity (Cabrera et al., 1990),

- The temperature and pressure of reactions in industrial applications (Rozanov and Treger, 2010, Wiberg and Motell, 1963 and Goldfinger et al., 1958).

The technology and the thermodynamic/kinetic data used in this case study are from the reports submitted by Rozanov and Treger (2010) and Goharrokhi et al. (2009), respectively.

Both methane and chlorine gases enter the PFR reactor at 25 °C and 1000 kPa. The reaction temperature is 430 °C and the CH_4/Cl_2 molar ratio is 0.5 (Fig. 1). The product is cooled down in a cooler resulting in a two phase flow, which is introduced to a flash separator. The vapor stream from the separator is non-product, while the liquid phase is considered as a product stream.

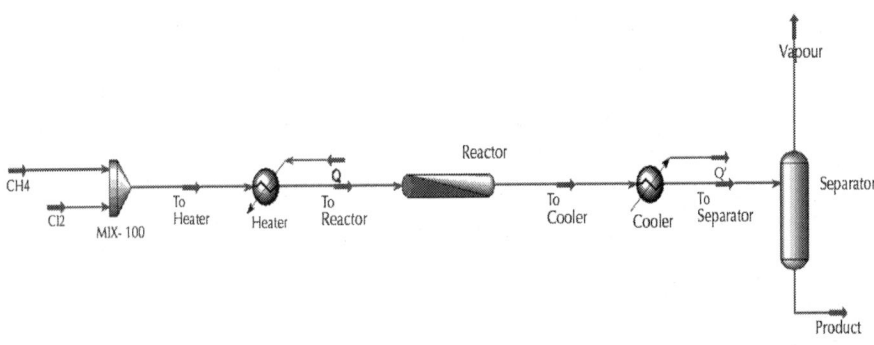

Figure 1: Chlorination of methane; original design without recycling.

Table 2 shows the compositions, the component flow rates in each stream and the total flow rate of streams in Fig. 1.

Table 2: Detailed information about the compositions and the flow rate of each stream in the design without recycling

Streams flow rates (kg/h)

Composition name	CH_4	Cl_2	T o heater	To reac-tor	To separa-tor	Vapor	Product	T o cooler
CH_4	802	0	802	802	203	188	15	203
CH_2Cl_2	0	0	0	0	300	2	298	300
$CHCl_3$	0	0	0	0	280	1	279	280
CCl_4	0	0	0	0	2790	3	2787	2790
Cl_2	0	7091	7091	7091	0	0	0	0
HCl	0	0	0	0	3646	1674	1972	3646
CH_3Cl	0	0	0	0	674	52	622	674
Total flow rate (kg/h)	802	7091	7893	7893	7893	1920	5973	7893

The estimated safety risks associated with each stream and based on the data of Table 1 is presented inTable 3.

Table 3: The estimated risk in affected people per year for each stream in design without recycling

Affected people/year								
CH$_4$	Cl$_2$	To heater	To reactor	To separator	Vapor	Product	To cooler
0	91.03	91.03	91.03	91.32	31.04	60.28	91.32

Eq. (2) estimates the total risk to the process based on Al-Sharrah et al. (2007) index, which is

$$K = 28,322 \text{ Affected People/Year}$$

While Eq. (5) calculates the total risk to the process based on the new proposed index, which is

$$(R.I)^T = 1196 \text{ Affected People/Year}$$

An alternative design to the chlorination process plant is to recycle the non-product (or waste) stream to the beginning of the process. Recycling is known as an effective and an economic way to increase the production rate and to minimize the wastes. In the present recycling design, the vapor stream from separator is pressurized by a compressor and then recycled to the plug flow reactor. Fig. 2 is a simplification of the chlorination process in order to illustrate the safety indices introduced earlier.

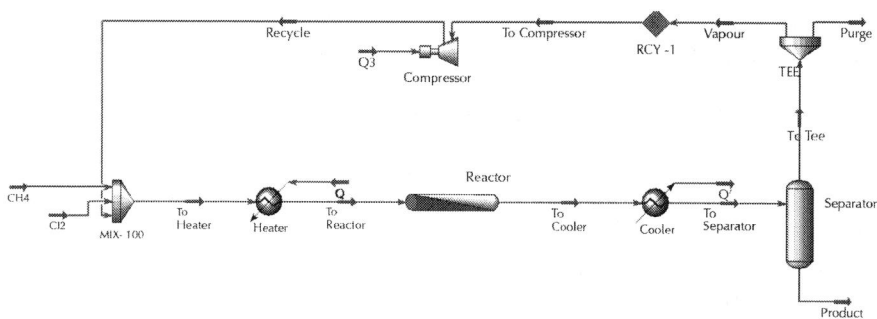

Figure 2: Chlorination of methane; alternative design with recycling.

The ratio of recycling stream to fresh feed streams is around 9 in this design. The non-product stream is connected to the compressor suction line resulting in putting the system under pressure. In case the compressor malfunctions, the pressure in the suction line might fluctuate. This is harmful to the process equipment and to the plant; in order to prevent any accident in the process a vacuum breaker is designed, which is called "purge" in this design.

So, the purge is used for equipment safety reasons and therefore, its flow rate in the steady state design is assumed to be zero. The product stream is subject to washing and purification in downstream process, which is not shown in Fig. 2. Table 4 provides data for the compositions and the flow rates of each stream in the alternative design.

Table 4: Detailed information about the compositions and the flow rate of each stream in the design with recycling

Streams flow rates (kg/h)												
Composition name	CH_4	Cl_2	To heater	To reactor	To separator	Vapor	Product	To cooler	Recycle	To compressor	To tee	Purge
CH_4	802	0	5084	5084	4314	4300	14	4314	4282	4282	4300	0
CH_2Cl_2	0	0	97	97	870	99	771	870	97	97	99	0
$CHCl_3$	0	0	63	63	1399	63	1336	1399	63	63	63	0
CCl_4	0	0	31	31	1083	30	1053	1083	31	31	30	0
Cl_2	0	7091	7091	7091	0	0	0	0	0	0	0	0
HCl	0	0	60,064	60,064	63,710	60,606	3104	63,710	60,064	60,064	60,606	0
CH_3Cl	0	0	1916	1916	2970	1946	1023	2970	1916	1916	1946	0
Total flow rate (kg/h)	**802**	**7091**	**74,345**	**74,345**	**74,345**	**67,045**	**7300**	**74,345**	**66,452**	**66,452**	**67,045**	**0**

Table 5 gives the estimated risks associated with each stream. Eq. (2) estimates the total risk to the process based on Al-Sharrah et al. (2007) index, which is

$K = 28, 322$ Affected People/Year

While the estimate based on the proposed index (Eq. 5) gives:

$(R.I)^T = 1196$ Affected People/Year

In both designs, it is assumed that the chlorine is dry and that the product (carbon tetrachloride) will be separated in downstream perfectly. A comparison is provided in Table 6.

Table 5: The estimated risk in affected people per year for each stream in the design with recycle

Affected people/year											
CH_4	Cl_2	To heater	To reactor	To separator	Vapor	Product	To cooler	Recycle	To compressor	To tee	Purge
0	91	1204.6	1204.6	1196.5	1123.7	72.8	1196.5	1113.6	1113.6	1123.7	0

Table 6: Comparison of the safety indices of the alternative designs

Risk indices (affected people/year)	Without recycling	With recycling
K	1641	28,322
$(R.I)^T$	91	1196

The objective of this process is to produce more carbon tetrachloride; therefore, by comparing the composition of the product stream in each design we can see which design meets this objective. For the same fresh feed rate to the PFR reactor (7893 kg/h), the original design produces more carbon tetrachloride (2787 kg/h) than the alternative design with recycle (1053 kg/h) as shown in Table 7.

Table 7: Comparison between the two designs for the CCl_4 production rate based on the same fresh feed rate to the reactor.

Composition	Product in original design (kg/h)	Product in alternative design (kg/h)
CH_4	15	14
CH_2Cl_2	298	771
$CHCl_3$	279	1336
CCl_4	2787	1053
Cl_2	0	0
HCl	1972	3104
CH_3Cl	622	1023
Total (kg/h)	5973	7300

Below is a list of remarks to explain why in spite of our anticipation, the original design is much safer than the alternative design:

- The size of the PFR reactor is the same for both designs.
- The fresh feed is the same for both designs.
- The recycling operation leads to mixing effects of the chemicals in PFR reactor and consequently, the PFR characteristics has changed to CSTR.
- The CSTR has usually lower conversion factor than PFR at the same volume due to the reasons mentioned in item c. That is why CCl_4 has lower production rate in recycling design (Table 7).

As indicated in Table 6, both K and $(R.I)^T$ indices estimate much higher risks for the recycling case compared to the original process. Therefore, considering the production rate of each process, the original design is preferred since the recycling of non-products has adverse effects on process safety as well as the production rate.

This case study shows that Eq. (2) always gives higher risks than Eq. (5) and therefore, it is misleading due to reasons mentioned in Section 3.

There are several explanations for such negative impacts of recycling on the process performance such as the selection of a wrong point for recycling, need for more cooling of product stream and so forth. This study showed that for this particular design, the recycling point was not a good option and therefore, it was rejected.

It has to be noted that total safety is different from the inherent safety of a process. Total safety is the summation of inherent safety and external safety of the process (Heikkilä et al., 1996 and Kletz, 2001). The new proposed index succeeds in taking into account both safety measures.

Case Study 2: Hydrogenation of Unsaturated Hydrocarbon

The hydrogenation of unsaturated hydrocarbon is an important process in oil and petrochemical industries for upgrading the hydrocarbons mixture. The best catalyst used for industrial applications is $Pd/\alpha\text{-}Al_2O_3$ due to its significant conversion factor and high selectivity effect (Krupka et al., 2006 and Seth et al., 2007).

Hydrogenation is a complex reaction, which takes place on a heterogeneous catalyst surface so; *Langmuir–Hinshelwood* rate expression is the best model for such a reaction (Evans and Wennerström, 1994). In a catalytic reaction two phenomena compete: diffusion and kinetics effects. They have influence on reaction rate. The smaller particle size of catalyst favors for kinetics phenomenon, which is higher rate (Zhou et al., 2007).

Ordouei et al. (2011) have designed a hydrogenation plant using the kinetic data listed in Table 8. The same data in the Table 8 have been used in this paper to present two design alternatives for hydrogenation process as shown in Figs. 3a, b and Figs. 4a, b.

Table 8: The kinetics data of hydrogenation reactions

Unsaturated hydrocarbon	k	E(J/mol)	Reference
1-Butene	1.482×10^{-5}	34,900	Seth et al. (2007)
Iso-butene	2.0958×10^{-6}	39,100	Seth et al. (2007)
Iso-octene	1.23×10^{-4}	10,506	Sarkar et al. (2006)
Styrene	0.0415	26,030	Zhou et al. (2007)

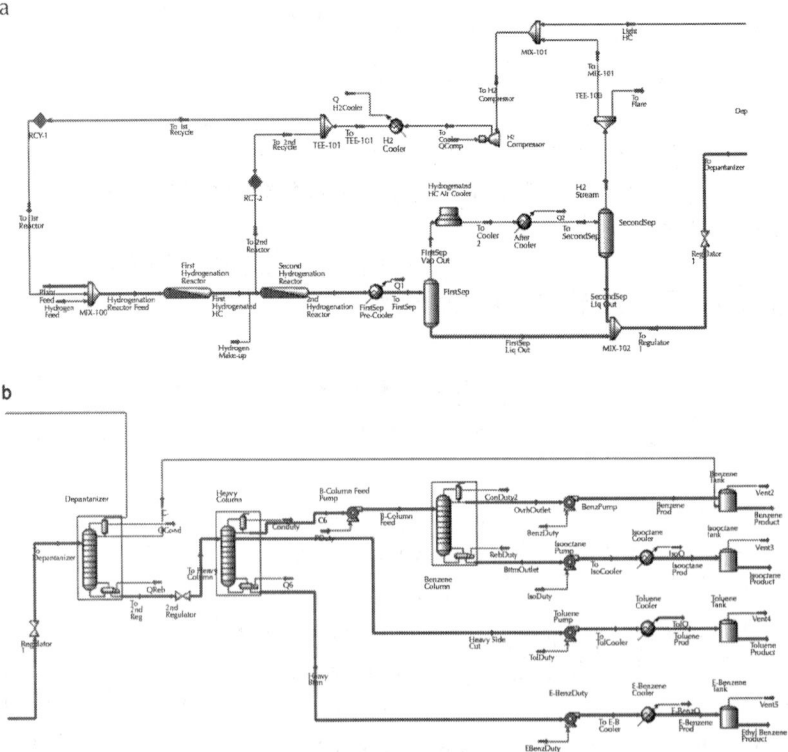

Figure 3: (a) Hydrogenation Process; original design: reaction and phase separation sections and (b) Hydrogenation Process; original design: purification section.

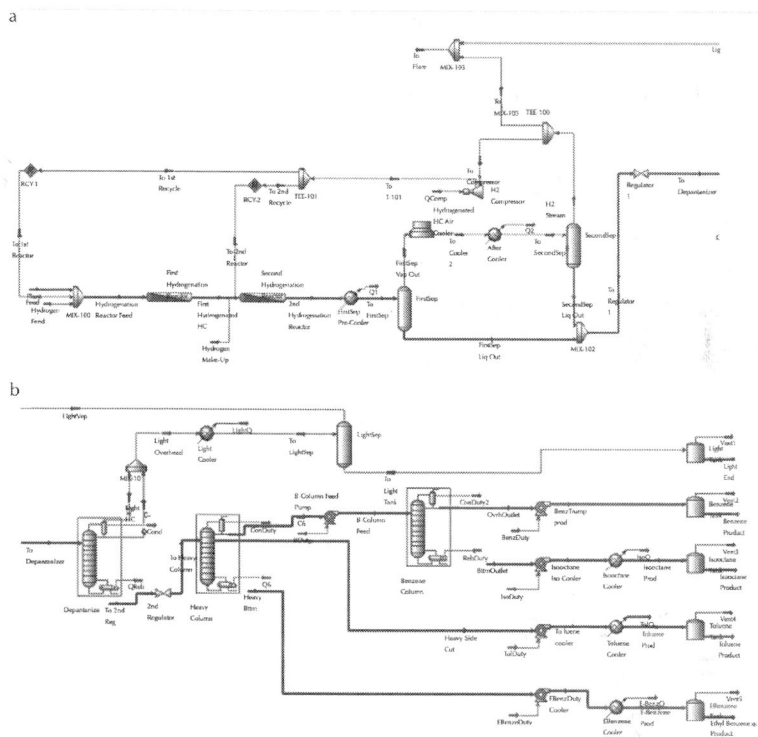

Figure 4: (a) Purification section: alternative design; reaction and phase separation sections and (b) Purification section: alternative design; purification section.

In Fig. 3a and b the plant feed contains unsaturated hydrocarbons mixture, which is mixed by hydrogen in the first adiabatic PFR reactor. The product of the PFR reactor is quenched by hydrogen in order to increase reaction efficiency in the second reactor. The product stream is then cooled in a shell and tube cooler and introduced to a flash drum, which separates vapor and liquid phases. The vapor phase is further cooled down in an air cooler followed by a shell and tube cooler before sending to a second flash drum.

The liquid from the first and the second flash drums are introduced to purification section. The main vapor portion from the second flash drum is pressurized in a compressor and returned to the beginning of the process and to the second PFR entrance for quenching purpose. The rest of the vapor and the vapor phase of the flash drum located on Depentanizer overhead are used in energy hub due to the heat content of the vapor stream.

From the Depentanizer, the light hydrocarbon bears more cooling followed by phase separation in a flash drum. The liquid outlet from this drum is then sent to Light End storage tank and the vapor phase is pressurized in a compressor and returned to the beginning of the process.

The bottom product of the Depentanizer goes to a Heavy Column. In heavy column ethyl benzene is separated from the bottom of the column and sent to the corresponding storage tank after being cooled down in a shell and tube cooler. From the column side cut toluene is isolated from the stream and sent to a cooler and then to a storage tank.

The Heavy Column overhead is sent to a Benzene Column where benzene is separated, cooled down and sent to corresponding storage tank. In this design, the pyrolysis gasoline is converted to five distinctive products compared to the alternative design, which generates four distinctive products

Unlike the original design, in the alternative design (Fig. 4a and b), the overhead and the side cut of Depentanizer are sent to the beginning of the process after being pressurized in a compressor.

The purity, flow rate, number of products and wastes, and the risk assessment of all products and waste streams in two process designs are summarized in Table 9.

The products and waste stream are listed in the first column. The table is divided into two sections "Original design" and "Alternative Design". Each part is divided into two sub-sections "Product Conditions" (divided into "Purity" and "Flow Rate") and "Risk Index" (divided into "Affected People per Year" and "Affected People per Ton Products"

Now, it is possible to analyze both processes in terms of inherent safer design based on the new developed indices.

Table 9: Summary of the risk analysis of hydrogenation original and alternative designs

| Products | Original design | | | | Alternative design | | | |
| | Products conditions | | Risk index | | Products conditions | | Risk index | |
	Flow rate, kg/h	Purity, %	#Affected people/year	#Affected people/ton Prodx10^{-6}	Flow rate, kg/h	Purity, %	#Affected people/year	#Affected people/ton Prodx10^{-6}
Light end	246.6	N/A	0.32	2.9	0		0	0
Benzene	6242	93	25.86	237.6	6111	87	24.46	223.7
Isooctane	760	57	2.76	25.4	648	18	2.62	23.9
Toluene	2325	71	5.96	54.8	2379	51	6.73	61.6
E-benzene	2853	70	6.18	56.8	3344	63	7.34	67.1
Total products	12427		41.08	377.5	12482		41.15	376.3
Total wastes	273.4		0.18	1.6	221.7		0.11	1.0

Table 10 gives the flow rates of the chemical species in all products and waste streams for both the original and the alternative designs. This data is obtained from HYSYS process simulations.

Table 10: The flow rates of chemical components in all streams for the original and the alternative designs from HYSYS simulations

Composition	Original design products, flow rates (kg/h)						Alternative design, products flow rates (kg/h)				
	Light ends	Benzene	Isooctane	Toluene	E-benzene	Wastes (H2 stream)	Benzene	Isooctane	Toluene	E-benzene	Wastes (H2 stream)
Iso-butane	79.4	1.0	0.0	0.0	0.0	48.2	91.0	0.0	0.0	0.0	38.4
n-Butane	62.2	25.7	0.0	0.0	0.0	26.5	99.0	0.0	0.0	0.0	8.1
Iso-pentane	0.0	117.9	0.0	0.0	0.0	2.0	117.2	0.0	0.0	0.0	2.9
n-Pentane	0.0	20.8	0.0	0.0	0.0	0.3	20.7	0.0	0.0	0.0	0.4
Iso-butene	71.6	3.8	0.0	0.0	0.0	37.0	97.5	0.0	0.0	0.0	16.2
Iso-octene	0.0	0.0	0.0	0.0	0.0	0.0	0.0	0.0	0.0	0.0	0.0
Iso-octane	0.0	335.3	498.9	554.0	0.9	0.7	261.4	154.8	966.2	6.4	1.0
1-Butene	33.4	2.4	0.0	0.0	0.0	16.7	52.8	0.0	0.0	0.0	7.4
Styrene	0.0	0.0	0.0	0.0	0.0	0.0	0.0	0.0	0.0	0.0	0.0
Ethyl-benzene	0.0	0.0	0.0	199.8	2067.7	0.0	0.0	0.0	169.1	2098.5	0.1
Benzene	0.0	5735.0	259.9	2.3	0.0	4.5	5371.9	493.9	128.6	0.1	7.7
Toluene	0.0	0.0	1.2	1569.3	784.3	0.3	0.0	0.0	1115.3	1239.2	0.5
Hydrogen	0.0	0.0	0.0	0.0	0.0	137.3	0.0	0.0	0.0	0.0	139.2
Total	246.6	6241.9	760.0	2325.3	2853.0	273.4	6111.5	648.7	2379.2	3344.2	221.7

Table 9 shows that the total flow rate and therefore, annual capacity of the alternative design is higher than that of the original design; however, the purity of all products in the original design is higher compared to the alternative design, which is absolutely a strong positive point due to producing more valuable products while using almost the same equipment and operation. Besides, there is one more product in the original design.

Therefore, Eq. (5) should be used to assess inherent safety, but in case of comparing two or more processes with significant differences in production rates, Eq. (6) should be used as for cross checking purposes. This fact shows the importance of Eq. (6) when process size (total annual capacity) is concerned.

In the above case study, Eq. (5) suffices to conclude that the original design is inherently safer due to similar production rates. However, Eq. (6) has been applied to the case study to probe the validity of above suggestion as the difference in normalized risk index is insignificant ($\sim$0.3%). And for such difference it is not wise to give-up the process with more products and higher purity.

In other words, one more product and less risk index of product streams, $(R.I)^P$, in the original design (41.08 max affected people per year) compared to the alternative design (41.15 max affected people per year) means that the original design is inherently safer. The difference between the normalized risk indices in maximum affected people per ton products (377.5×10^{-6} for the original design and 376.3×10^{-6} for alternative design) is about 0.3% and negligible.

The risk index for individual product streams in one design may be more or less compared to another design, but the overall risk index for the whole product streams within the process does matter. For instance, from Table 9, the annual capacity of benzene production and its purity in the original design are higher than those in the alternative design by 1100 tons/year and 6%, respectively. Therefore, the associated risk indices are higher.

Also, there is no linear dependency between the flow rates of two distinctive streams and the associated risk indices within the same process; for example, the flow rate of benzene stream is 2.5 times as much as the toluene stream and about twice compared to ethyl-benzene in the original design; however, the risk indices of the benzene stream are more than four times as much as both toluene and ethyl-benzene product streams.

Eq. (5) divulges that the risk associated with a process design is highly dependent on the mass fraction of hazardous impurities in the product streams. The state of the art of inherently safer design is to reduce the hazardous chemicals in the product streams. And the most convenient step for economically source (hazard) reduction is the conceptual design (Tchobanoglous, 2009).

For the waste stream, even though the risk indices of waste stream in the original design are higher compared to the alternative design, they are about 0.4% of the total risks in corresponding design [Eqs.(5) and (6)] and again it is negligible.

Moreover, the so-called waste stream in our case study has heat content and can be used for heat generation, as stated before. So, it may be considered as a byproduct stream.

To recap, the total risks based on Eq. (5) are 41.26 for both designs and based on Eq. (6) are 379.1×10^{-6} for the original design and 377.3×10^{-6} for the alternative design.

Thus, the original design is more desirable due to diversity of products, higher purity of products and inherently safer design.

CONCLUSIONS

This paper offers new simple inherent safety indices, which can be employed during primitive stage of a process design, i.e. the conceptual design, to eliminate or minimize the hazardous chemicals in the process involved. The new indices have the following advantages and contributions on inherently safer process design (ISD):

- Simple and user friendly
- Quantitative and mathematical model for risk evaluation

- Safe results and based on the reliable database of undesirable events or accidents
- Predicts the risks of a process in terms of human fatalities
- The metrics "frequency of accidents", "severity of accident" and "chemical inventory" have great influences on the risks associated with a process. The new indices probe that the mass fraction of hazardous chemicals in process streams has also substantial contribution to hazardousness of a process.
- A linear relationship between the risk index and the mass fraction of hazardous materials has been established in this paper.
- They can be used as a strong screening tool for design engineers, decision makers and regulators.
- They can be used as a commercial tool for publicity to convince the regulators and potential clients for their commitment to safe design.
- Since the most convenient step for source/hazards reduction is the conceptual design, the new indices help to reduce capital costs by a decrease in hazards generation leading to design smaller waste treatment facilities and control stations. Hence, the core of the presented new indices in this paper is economically and inherently safer process design.

REFERENCES

1. Al-Sharrah, G.K., Hakinson, G., Elkamel, A., 2006. Decision making for petrochemical planning using multiobjective and strategic tools. Trans. IChemE, Chem. Eng. Res. Des. 84 (A11), 1019–1030.

2. Al-Sharrah, G.K., Edwards, D., Hakinson, G., 2007. A new safety risk index for use in petrochemical planning. Trans. IChemE 85 (B6), 533–540.

3. Al-Sharrah, G., Elkamel, A., Almanssoor, A., 2010. Sustainability indicators for decision making and optimization

in the process industry: the case of the petrochemical industry. Chem. Eng. Sci. 65 (4), 1452–1461.

4.   ARIP (Accidental Release Information Program), 1999. ⟨http://www.epa.gov/ osweroe1/tools.htm#arip⟩ (November 29, 2011).

5.   Belke, J.C., 2000. Chemical Accident Risk in U.S. Industry – A Preliminary Analysis of Accident Risk Data from U.S. Hazardous Chemical Facilities. United States Environmental Protection Agency, Washington, DC, USA.

6.   Bucsi, I., Olah, G.A., 1992. Selective monochlorination of methane over solid acid and zeolite catalysts. Catal. Lett. 16, 27–38.

7.   Cabrera, M.I., Alfano, O.M., Cassano, A.E., 1990. Product yield and selectivity studies in photoreaction design, theory, and experiments for the chlorination of methane. Chem. Eng. Sci. 45 (8), 2439–2446.

8.   Canadian Centre for Occupational Health and Safety, 2009. Hazard Risk. ⟨http:// www.ccohs.ca/oshanswers/hsprograms/ hazard_risk.html⟩ (February 28, 2012).

9.   Carson, P., Mumford, C., 2002. Hazardous Chemicals Handbook, 2nd ed. Butterworth Heinemann; Great Britain.

10.   CCCP, 2009. Inherently Safer Chemical Processes: A Life Cycle Approach, 2nd ed. John Wiley & Sons, Hoboken, NJ. Cordella, M., Tugnoli, A., Barontini, F., Spadoni, G., Cozzani, V., 2009. Inherently safety of substances: identification of accidental scenarios due to decomposition products. J. Loss Prev. Process Ind. 22, 455–462.

11.   Couper, J.M., Penny, W.R., Fair, J.R., Walas, S.M., 2005. Chemical Process Equipment Selection and Design, 2nd ed. Elsevier; United States of America (Chapter 0: Rules of Thumb Reactors; Storage Tanks).

12.   Denisov, E.T., 1974. Liquid-Phase Reaction Rate Constants (R.K. Johnston Trans.). IFI/ Plenum (Chapter VI).

13.   EPA, 2012. US Environmental Protection Agency's Website, Chemical Process Simulation for Waste Reduction. WAR

algorithm. ⟨http://www.epa.gov/nrmrl/ std/war/sim_war.htm⟩ (May 28, 2012).

14. Evans, D.F., Wennerström, H., 1994. The Colloidal Domain: Where Physics, Chemistry, Biology and Technology Meet. VCH Publishers, Inc.; Weinhein, Germany (Chapter 2).

15. Goharrokhi, M., Torabi, M., Akbari, F., Colozari, F., 2009. An introduction to simulation and optimization of steady state chemical processes with ASPEN PLUS 2006. Daneshgaran Sanat Pajouh Co., Nasim Distributor Center.

16. Goldfinger, P., Jeunehomme, M., Meartens, G., 1958. Energies and entropies of activation in chlorine atom reactions. J. Chem. Phys. 29, 456–458.

17. Heikkilä, A.M., Hurme, M., Järvelainen, M., 1996. Safety considerations in process synthesis. Comput. Chem. Eng. 20, S115–S120.

18. Heikkilä, A.M., 1999. Inherent Safety in Process Plant Design. VTT Publication. 384. Technical Research Centre of Finland (VTT), Vuorimiehentie, Finland.

19. Hendershot, D.C., 2011a. Inherently Safer Design: An Overview of Key Elements. Professional Safety. file:///I:/Papers/Safety%20Paper/Sent%20to%20Publisher/ Chemical%20 Engineering%20Science/048_055_F2Hendershot_0211Z. pdf.

20. Hendershot, D.C., 2011b. Inherently safer design – not only about reducing consequences. Process Saf. Prog. (AIChE) 30 (4), 351–355. Henselwood, F., Phillips, G., 2006. A matrix-based risk assessment approach for addressing linear hazards such as pipelines. J. Loss Prev. Process Ind. 19 (5), 433–441.

21. Khan, F.I., Abbasi, S.A., 1998. Inherently safer design based on rapid risk analysis. J. Loss Prev. Process Ind. 11, 361–372.

22. Khan, F.I., Amyotte, P.R., 2002. Inherent safety in off-shore oil and gas activities: a review of the present status and future directions. J. Loss Prev. Process Ind. 15, 279–289.

23. Kletz, T., 2001. Learning from Accidents, 3rd ed. Gulf Professionals Publishing (Chapter 30). Koller, G., Fischer, U.,

Hungerbühler, K., 1999. Assessment of environment, health and safety aspects of fine chemical processes during early design phase. Comput. Chem. Eng. (Suppl.), S63–S66.

24. Koller, G., Fischer, U., Hungerbühler, K., 2001. Comparison of methods suitable for assessing the hazard potential of chemical processes during early design phases. Trans. IChemE 79 (Part B), 157–166.

25. Krupka, J., Severa, Z., Pasek, J., 2006. Competitive hydrogenation in binary diene systems on a palladium catalyst. React. Kinet. Catal. Lett. 89 (2), 359–368.

26. Lees, F.P., 1996. Loss Prevention in the Process Industries: Hazard Identification, Assessment and Control, 2nd ed.vol. 1. Butterworth-Heinemann. Leong, C.T., Shariff, A.M., 2008. Inherent safety index module (ISIM) to assess inherent safety level during preliminary design stage. Process Saf. Environ. Prot. 86, 113–119.

27. Levenspeil, O., 1999. Chemical Reaction Engineering, 3rd ed. John Wiley & Sons; United States of America (Chapter 10).

28. Lewin, D.R., 2004. Design and Analysis (054402), Lecture 10: Interaction of Process Design and Control, ⟨http://tx.technion. ac

29. Marhavilas, P.K., Koulouriotis, D.E., 2008. A risk estimation methodological framework using quantitative assessment techniques and real accidents' data: application in an aluminum extrusion industry. J. Loss Prev. Process Ind. 21 (6), 596–603.

30. Marhavilas, P.K., Koulouriotis, D., Gemeni, V., 2011. Risk analysis and assessment methodologies in the work sites: on a review, classification and comparative study of the scientific literature of the period 2000–2009. J. Loss Prev. Process Ind. 24, 477–523.

31. Missen, R.W., Mims, C.A., Saville, B.A., 1999. Introduction to Chemical Reaction Engineering and Kinetics. John Wiley & Sons; United States of America (Chapter 5).

32. Ordouei, M.H., Biglari, M., Mujiburohman, M., 2014. A novel process design and simulation for hydrogenation plant in refineries. Energy Source, Part A: Recover. Util. Environ. Eff. (in press).

33. Pankratz, T.M., 2001. Environmental Engineering Dictionary and directory. Lewis Publishers; United States of America (p. 271).

34. Reniers, G.L.L., Dullaert, W., Ale, B.J.M., Soudan, K., 2005. Developing an external domino prevention framework: Hazwim. J. Loss Prev. Process Ind. 18, 127–138.

35. Rozanov, V.N., Treger, Y.A., 2010. Kinetics of the gas-phase thermal chlorination of methane. Kinet. Catal. 51 (5), 635–643.

36. Sarkar, A., Seth, D., Ng Flora, T.T., Rempel, G.L., 2006. Kinetics of liquid-phase hydrogenation of isooctenes on a Pd/γ-alumina catalyst. Am. Inst. Chem. Eng. J. 52 (3), 1142–1156.

37. Seth, D., Sarkar, A., Ng, F.T.T., Rempel, G.L., 2007. Selective hydrogenation of 1,3 butadiene in mixture with isobutene on a Pd/α-alumina catalyst in a semibatch reactor. Chem. Eng. Sci. 62 (17), 4544–4557.

38. Shariff, A.M., Rusli, R., Leong, C.T., Radhakrishnan, V.R., Buang, A., 2006. Inherent safety tool for explosion consequences study. J. Loss Prev. Process Ind. 19, 409–418.

39. Shariff, A.M., Leong, C.T., 2009. Inherent risk assessment – a new concept to evaluate risk in preliminary design stage. Process Saf. Environ. Prot. 87, 371–376.

40. Tchobanoglous, G., 2009. Environmental Engineering, Environmental Health and Safety for Municipal Infrastructure, Land Use and Planning, and Industry. In: Nemerow N. L., Agardy F.J., Sullivan P. and Salvato J. A. (Eds.), (Chapter 3: Solid Waste Management), 6th ed. John Wiley & Sons; United States of America.

41. Tixier, J., Dusserre, G., Salvi, O., Gaston, D., 2002. Review of 62 risk analysis methodologies of industrial plants. J. Loss Prev. Process Ind. 15, 291–303.

42. Wiberg, K.B., Motell, E.L., 1963. The kinetic isotope effect in the photo-chlorination of methane. Tetrahedron 19, 2009–2023.

43. Woodruff, J.M., 2005. Consequence and likelihood in risk estimation: a matter of balance in UK health and safety risk assessment practice. Saf. Sci. 43, 345–353.

44. Zhou, Z.M., Cheng, Z.M., Cao, Y.N., Zhang, J.C., Yang, D., Yuan, W.K., 2007. Kinetics of the Selective Hydrogenation of Pyrolysis Gasoline. Chem. Eng. Technol. 30 (1), 105–111.

Chapter 8

Perspectives on Chemical Hazard Characterization and Analysis Process at DOE

J.C. Laul[1], Fred Simmons[2], James E. Goss[3], Lydia M. Boada-Clista[4], Robert D. Vrooman[5], Rodger L. Dickey[6], Shawn W. Spivey[7], Tim Stirrup[8], and Wayne Davis[9]

[1]Ph.D, CHMM, REM, is affiliated with Safety Basis Division, Los Alamos National Laboratory, MS K489, Los Alamos, NM 87545, USA

[2]Affiliated with Washington Savannah River Company, 731-N, Aiken, SC 29808, USA

[3]PE, CSP, is affiliated with National Nuclear Security Administration, Y-12 Site Office, Oak Ridge, TN 37831, USA

[4]Affiliated with Ohio Field Office, 175 Tri-County Parkway, Springdale, OH 45246, USA

complete an EPHA. However, there are also additional features of the EPHA that go beyond the scope of DSAs and PrHAs.

In addition to the benefits that this report may provide to the emergency management program, some parts of the SB process may also benefit explosive facilities (29 CFR 1910.109) that are required to complete a process hazards analysis under the Occupational Safety and Health Administration (OSHA) PSM requirement (29 CFR 1910.119). "RELATED TOPICS" discusses these topics in more detail. A non-nuclear facility referred to here may be a radiological facility, with below Hazard Category 3 quantities as defined in DOE-STD-1027; facilities that use or store explosives, accelerators, facilities that use or store hazardous chemicals, laboratories, biological research facilities, and general industrial type facilities.

APPLICABILITY

This report presents a proposed methodology that may be used for non-nuclear facilities or sites that have chemicals present that may represent a hazard to the worker, the environment, or the public. This report is not intended for facilities with only incidental or standard industrial usage of chemicals, such as the use of cleaning products in an office area. Note that this report is *not* a proposed standard or guidance for a Safety Basis (SB) process or safety document. This report simply outlines various SB steps and methodologies involved and their advantages and disadvantages associated with them, so that each site can decide on its own the merits and demerits of each approach. Adoption of any step of the SB process is voluntarily.

RELEVANT GUIDANCE, REGULATIONS, AND DOE ORDERS

The United States (U.S.) Department of Energy (DOE) has an ISMS policy that requires hazard analyses and implementation of controls to protect the workers and public. The ISMS applies to all

DOE facilities as required by DOE-P-450.4, *Safety Management System Policy*, and DEAR clause 48 CFR 970.5223-1,*Integration of Environment, Safety, and Health into Work Planning and Execution*. The DEAR clause requires DOE contractors to integrate environment, safety, and health into work planning and execution with guiding principles. The ISMS is further supported with additional relevant regulations and DOE orders.

Safety Basis for Hazard Category 1, 2, and 3 nuclear facilities is required by 10 CFR 830, Subpart B, *Safety Basis Requirements*. Previously, DOE required through DOE-O-5481.1B that non-nuclear facilities also develop SB documentation. Now, in essence, the ISMS DEAR clause provides the overarching safety basis requirements for non-nuclear facilities. There is not a DOE order or standard that provides specific guidance for the development of SB documentation for non-nuclear facilities. Yet, there may be an expectation that non-nuclear facilities should also develop SB documentation for non-nuclear facilities. Various DOE sites over the years have adopted site-specific requirements for chemical SB processes and documentation resulting in wide variations across the DOE complex.

In addition to the ISMS, other DOE regulations, and Occupational Safety and Health Administration (OSHA), or U.S. Environmental Protection Agency (EPA) regulatory requirements may apply to non-nuclear facilities. Some sites take the position that hazards with existing Federal regulations and consensus standards are not unique hazards and already have sufficient controls identified and that the challenge is to consistently apply the controls. The following is a listing of requirements that have parts that can be related to the SB process or elements of the SB process. The list was adapted from DOE-HDBK-1163-2003, *Integration of Multiple Hazard Analysis Requirements and Activities* (Hazard Analysis Handbook).

- 10 CFR 830, *Subpart B, SB Requirements*
- 10 CFR 850, *Chronic Beryllium Disease Prevention Program*
- 29 CFR 1910 and 1926, *Various Hazard or Activity Specific OSHA regulations*

- 29 CFR 1910.109, *Explosives and Blasting Agents*
- 29 CFR 1910.119 and 29 CFR1926.64, *Process Safety Management*
- 29 CFR 1910.120, *Hazardous Waste Operations and Emergency Response*
- 40 CFR 302.4, *Designation, Reportable Quantities, and Notification*
- 40 CFR 355, *Emergency Planning and Notification*
- 40 CFR 372, *Toxic Chemical Release Reporting: Community Right to Know (This regulation is not directly related to SB, but is useful for reporting requirements)*
- 40 CFR 68, *Chemical Accident Prevention Provisions*
- 48 CFR 970.5204-2 (c)(2), *Laws, Regulations, and DOE Directives*
- DOE-G-151.1-1 V2, *Hazards Surveys and Hazards Assessments*
- DOE-G-420.1-2, *Guide for the Mitigation of Natural Phenomena Hazards for DOE Nuclear Facilities and Non-nuclear Facilities*
- DOE-M-440.1-1, *Explosives Safety Manual*
- DOE-O-151.1C, *Comprehensive Emergency Management System*
- DOE-O-420.1A, *Facility Safety*
- DOE-O-440.1A, *Worker Protection Management*
- DOE-P-450.4, *Safety Management System Policy*
- DOE-G-440.1, Locally enforced fire/building codes

This listing is not intended to imply that the requirements specifically drive the DOE-based SB process. Several of these regulations could be used as the basis for the SB process or development of safety document for non-nuclear facilities, such as *Chemical Accident Prevention Program (40 CFR 68) and Process Safety Management (29 CFR 1910.119)*. In addition, this list identifies multiple requirements for hazards analysis, and combining these multiple requirements into a single effort could minimize SB efforts as suggested by the Hazards Analysis Handbook (DOE-HDBK-1163-2003, October 2003). Appendix A provides description of these CFR regulations and DOE orders.

A POSSIBLE SAFETY BASIS PROCESS

In compliance with the guiding principles of the DEAR clause and ISMS, there are some six main steps in developing SB documentation which include the essential features of the five core functions of the ISMS, as shown in Figure 1. The SB documentation development is an iterative process and can be developed using a graded approach. The key steps are as follows:

- **Facility and Work Description:** Describe the facility and define the work to be performed.

- **Hazard Identification:** Identify hazards (e.g., chemical, physical, electrical, industrial) and potential initiators that could lead to an accident.

- **Facility Chemical Hazard Classification (CHC):** Performing a facility CHC is not required by the ISMS. However, it is an optional, useful step in the SB process. A facility CHC can be described in the facility and work description or hazard identification section or it can be a stand alone section.

- **Hazard Analysis (HA):** Perform hazard analysis that can be qualitative or quantitative depending on the nature of hazard and hazard facility (i.e., High, Moderate, and Low).

- Qualitative HA is discussed using industry approach and DOE-STD-3009 nuclear facility-like approach, and various hazard analyses methodologies are discussed.

- Quantitative HA (consequence/source term analysis) is discussed using various applicable chemical dispersion models.

- **Identification of Controls:** Develop hazard controls (e.g., engineered, administrative) to eliminate, limit, or mitigate identified hazards and to protect the workers and public. Define the process(es) for maintaining hazard controls.

- **Commitments to Safety Management Program:** Define commitments in terms of maintaining controls to perform work safely and ensure safe performance and operation of the facility.

- **Document and Approval Process:** Prepare SB documentation or safety document using the above steps. Approval is usually required or negotiated between the contractor and the field or site office of DOE/NNSA, depending on the level of the chemical hazard in the facility.

The details of each of the above steps are provided in the following sections.

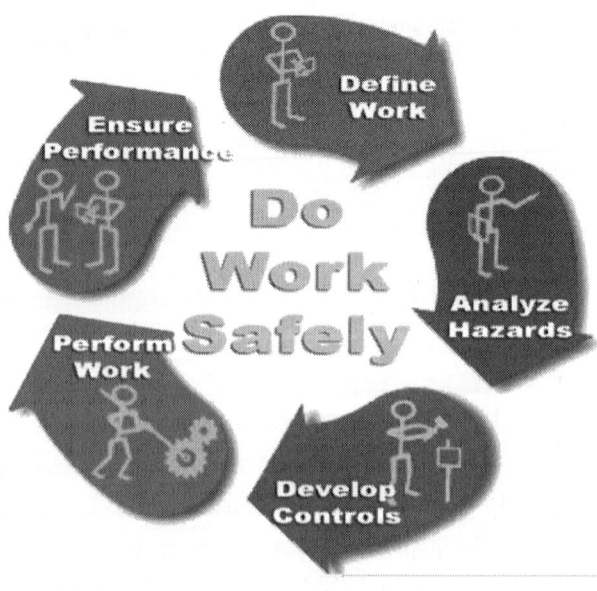

Figure 1: Five Core Steps of the ISMS (LANL, Laul 2001[2]; Cournoyer & Maestas 2004[3]).

FACILITY AND WORK DESCRIP-TION

A thorough description of the facility, the chemical process system, and associated work activities being assessed are provided in this initial step. The description typically should include the site where the facility is located, the facility identification (e.g., building number and location), building configuration, and principal activities performed inside the facility. Site and facility description is provided to aid understanding of potential hazardous materials and operations.

These descriptions focus on facility features and work processes necessary to understand the hazard analysis and accident analysis, not just those structures, systems, and components (SSCs) important to safety. The descriptions may provide the following types of information.

- Overview of the facility, material inputs/outputs, mission, and history;
- Description of the facility structure and design basis;
- Description of the facility process systems and constituent components, instrumentation, controls, operating parameters, and relationships of SSCs;
- Description of bulk storage location and confinement systems;
- Description of the facility safety support systems;
- Description of the facility utilities, with schematic outline of the basic utility distribution systems;
- Description of individual processes within the facility.

The description of the individual process may include details on basic process parameters, summary of types and quantities of hazardous materials, process equipment, instrumentation and control systems and equipment, basic process flow diagrams, piping & instrumentation documents (P&IDs), and operational considerations associated with individual processes or the facility. Existing supporting documentation may be referenced.

impacts. However, these regulations provide an initial practical method to identify more generally recognized hazards. Readers may consult each enabling regulation for a better understanding for the purpose and objective.

29 CFR 1910.119 and 29 CFR 1926.64, Process Safety Management

This OSHA regulation currently lists 137 chemicals and their threshold quantities (TQs). It also includes all flammable liquids and gases with a TQ of 10,000 lbs with a couple of exceptions for liquids. Typically these TQs are used to determine when industry is required to perform an in-depth analysis of the process (e.g., PrHA) to ensure the safety of the workers. Using these TQs to screen for chemicals that could be considered a hazard has advantages and disadvantages, which are as follows.

Advantages

- Using these TQs brings the DOE facility in line with requirements for private industry to perform special analyses when these limits are exceeded for a facility.
- Using these TQs is a simple and fast method for determining when hazardous quantities of specific chemicals that should be further analyzed are present.
- Using these TQs provides a list of chemicals that could be hazardous from many different perspectives (e.g., toxic, flammable, explosive, or corrosive).
- Processes and hazards for analyzing these chemicals are available from private industry to aid in any analysis.
- Using these TQs may enable a facility to impose limit on quantities of the chemical to below TQ levels and thus be exempt from this regulation (i.e., no need for a PrHA).

Disadvantages

- There are only 137 chemicals listed in this regulation, plus flammable liquids and gases with few exceptions for liquids. The vast majority of chemicals in DOE or private industry accidents are not listed on this list. For example sulfuric acid is only represented by oleum (fuming sulfuric acid).

- The list in this regulation does not correspond with lists obtained from other regulations. Therefore, a danger of improper overlap occurs when this regulation is used in conjunction with other regulations.

- Quantities of chemicals listed in this regulation could be much greater than that necessary to cause a severe accident. For example, the limit for ammonium perchlorate, which is either shock-sensitive or a Class 4 oxidizer depending on the particle size, is 7,500 pounds.

- Concentration thresholds are supplied for some chemicals. For example, for nitric acid at 94.5% and above the limit is 500 pounds. However, if the concentration of nitric is below 94.5% then this regulation does not apply.

- Reactive chemistry is not commonly addressed in PSM-listed chemicals and thus a PrHA should include chemical reactive hazards also, where possible.

40 CFR 68, Chemical Accident Prevention Provisions

This EPA regulation establishes a list of 140 regulated substances and their TQs for stationary sources concerning the prevention of accidental releases to protect the public. It further establishes a list of toxic endpoints for offsite consequence analysis and sets the requirements for a Risk Management Plan (RMP) if TQs are exceeded. It is to note that only 40% of the RMP listed chemicals overlap with the PSM listed chemicals. The TQs for the RMP chemicals are usually higher than TQs for the PSM chemicals;

because the RMP chemical process focuses towards to protecting the public, while the PSM chemical process focuses towards to protecting the worker.

Advantages

- Using these TQs is a simple and fast method for determining the presence of hazardous quantities of specific chemicals that should be further analyzed.
- Using TQ values puts in place requirements that are triggered by Federal requirements.
- Using these TQs may enable a facility to impose limit on quantities of the chemical to below TQ levels and thus be exempt from this regulation (i.e., no need for a PrHA).

Disadvantages

- Chemicals under this regulation are listed due to their health hazards or flammability. The list is limited to 77 toxic and 63 flammable substances, for a total of 140. The vast majority of chemicals in DOE or private industry accidents are not listed on this list. For example, HCN (hydrogen cyanide) is not on the list but is highly toxic.
- Chemicals listed are not consistent with chemicals listed in other enabling regulations. For example, ammonia in solution has a 29 CFR 1910.119 (PSM) threshold quantity (TQ) of 15,000 pounds (44% solution) and a 40 CFR 302 reportable quantity (RQ) of 100 pounds and a 40 CFR 355 threshold planning quantity (TPQ) of 500 pounds (10% solution) and a RMP TQ of 20,000 pounds (20% solution).
- Reactive chemistry is not commonly addressed in RMP listed chemicals and thus a PrHA should also include chemical reactive hazards, where possible.
- There are many provisions in this regulation that could become confusing if used in a SB process, especially if this regulation

is used in conjunction with other regulations. First, the three levels of reporting alluded to in this section are dependent upon both the product being present in quantities greater than a TQ and if an accident with the product had occurred within the previous five years. Second, this regulation is based upon a list of chemicals and their threshold quantities that would trigger the need to meet this regulation. This list of chemicals is of 140 items and does not coincide with other lists such as that found for the PSM standard (e.g., 40% overlap).

- One area where these lists do not coincide is in TQs. For example, the TQ for arsine in the PSM standard is 100 pounds while the TQ in this RMP standard is 1,000 pounds. There are other such examples (see table in "40 CFR 355, Emergency Planning and Notification").

40 CFR 355, Emergency Planning and Notification

This EPA regulation establishes the list of extremely hazardous substances, TPQ, and facility notification responsibilities necessary for the development and implementation of State and local emergency response plans. Chemicals are listed with an RQ and a TPQ value. Those chemicals not appearing on the list have an RQ and a TPQ of 10,000 pounds by default.

Advantages

- Using these RQs and TPQs is a simple and fast method for determining when hazardous quantities of specific chemicals that should be further analyzed are present.
- One can choose whether RQs or TPQs are used in the screening process.
- Using RQ and TPQ values puts in place requirements that are triggered by Federal requirements.

Disadvantages

- Chemicals listed in this regulation are listed due to their health hazards. Chemicals with other hazards (e.g., Na and K) are not listed, and are thus automatically defaulted to the 10,000-pound limit while a similar material with respect to reactivity, but not toxicity, phosphorous pentachloride (PCl_5), has an RQ and TPQ of 500 pounds.
- Chemicals listed are not consistent with chemicals listed in other regulations. For example anhydrous ammonia gas has a 29 CFR 1910.119 (PSM) TQ of 10,000 pounds and 40 CFR 302 RQ of 100 pounds and a 40 CFR 355 TPQ of 500 pounds.
- Chemical RQ and TPQ values are not consistent with screening values from other regulations.
- RQ and TPQ values from this list vary from being the same to having a 500-fold difference, which can cause confusion.

40 CFR 302.4, Designation, Reportable Quantities, and Notification

This EPA regulation identifies RQs for a list of hazardous substances, and sets forth the notification requirements for releases of these substances. This regulation also establishes reportable quantities for hazardous substances designated in the Clean Water Act (CWA).

Advantages

- Provides a detailed list of substances with regulatory limits.
- A fast way of identifying the relative risk of a reportable release vs. the amount of a substance in a facility.

Disadvantages

- Inconsistent use of chemical nomenclature when compared to lists supplied in other regulations as shown below.

- Using this list and associated quantities in a process can become confusing. The list in this regulation does not coincide with lists from other regulations such as PSM or 40 CFR 68 (RMP). Items that are on this list may not be present on other lists. Hazardous materials such as arsine are listed on the PSM list and the list from 40 CFR 68 but are absent from this regulation.

- Likewise, hazardous materials on this list may not be found on any other list. Another difficulty is that RQs and TQs from the various lists do not coincide and there is no relationship between the RQ and TQ values from these lists.

- As with other regulations listed above, there could be difficulties if the list of extremely hazardous substances (EHSs), RQs TPQs, and TQs is used in the SB process. This difficulty stems from inconsistencies between those items listed in these various lists and differences between listed quantities. For examples, see the table below.

Chemical	29 CFR 1910.119 TQ (lbs)	40 CFR 68 TQ (lbs)	40 CFR 302 RQ (lbs)	40 CFR 355 TPQ (lbs)
Arsine	100	1,000	–	100
Fluorine	1,000	1,000	10	500
Methyl isocyanate	250	10,000	10	500
Hydrogen chloride	5,000	5,000	5,000	500

As can be seen in this table there is no relationship between these lists or the various quantities listed. In some cases (e.g., arsine, methyl isocyanate) values for 29 CFR 1910.119 are 10 to 40 times less than 40 CFR 68, while in other cases they are the same (hydrogen chloride, fluorine). Values from 40 CFR 302 range from being equal to 29 CFR 1910.119 and 40 CFR 68 (RMP) or much greater than 40 CFR 355 (hydrogen chloride) to being up to 50-fold less than 40 CFR 355 values and up to 1,000 times less than 40 CFR 68 (methyl isocyanate). This table shows how the use of these inventory-based regulations by themselves could lead to some confusion, and thus requires careful consideration and integration.

GENERIC DOE ORDERS

DOE-O-420.1A, Facility Safety

This DOE order establishes facility safety requirements related to nuclear safety design, criticality safety, fire protection, and NPHs mitigation. Portions of this order apply to non-nuclear facilities.

Advantage

- Familiarity with nuclear safety documentation makes it relatively easy to develop a plan for non-nuclear facility. This order provides requirements and criteria for assessing fire and NPH.

Disadvantage

- For a non-nuclear facility, only two types of hazards are addressed (e.g., fire and NPH) and this order lacks guidance on a graded approach.

DOE–O- 440.1A, Worker Protection Management

This DOE order establishes the framework for an effective worker protection program that will reduce or prevent injuries, illnesses, and accidental losses by providing DOE Federal and contractor workers with a safe and healthful workplace.

Advantages

- Provides a list of codes and standards to follow.
- Provides a detailed list of requirements beyond the code.

Disadvantage

- Does very little to assist in identifying the hazard except to reference the codes and standards.
- Single Chemical Regulations

10 CFR 850, Chronic Beryllium Disease Prevention Program

This health and safety regulation establishes a chronic beryllium disease prevention program (CBDPP) that supplements and is integrated into existing worker protection programs that are established for DOE employees and DOE contractor employees.

Advantages

- Hazard identification is simple, "Is beryllium present?"
- Not applicable, if beryllium is not present.

Disadvantage

- Some of the requirements are vague, leading to inconsistent implementation. For example, sampling for beryllium is required, however, the sampling technique, which can dramatically affect detection limits and results, is not specified. On the other hand, toxicity and dose/exposure are independent of detection limits.

Chemical-specific OSHA Regulations as Found in 29 CFR 1910 and 29 CFR 1926

There are many chemicals that have specific OSHA regulations as found in 29 CFR 1910 and 29 CFR 1926. The 1910 refers to facility operation and 1926 refers to construction. While many chemicals overlap between 1910 and 1926, only one regulation is cited for those chemicals. These are shown below.

1910.1001 – Asbestos	1910.1002 – Coal tar pitch volatiles
1910.1003 – 13 carcinogens (4-nitrobiphenyl, etc.)	1910.1004 – Alpha-naph-thylamine
1910.1006 – Methyl chloro-methyl ether	1910.1007 – 3,3'-Dichloro-benzidine (and its salts)
1910.1008 – bis-Chloromethyl ether	1910.1009 – Beta-naphthyl-amine
1910.1010 – Benzidine	1910.1011 – 4-Aminodiphe-nyl
1910.1012 – Ethyleneimine	1910.1013 – Beta-propiolac-tone
1910.1014 – 2-Acetylamino-fluorene	1910.1015 – 4-Dimethylami-noazobenzene
1910.1016 – N-Nitrosodimeth-ylamine	1910.1017 – Vinyl chloride
1910.1018 – Inorganic arsenic	1910.1025 – Lead
1910.1027 – Cadmium	1910.1028 – Benzene
1910.1029 – Coke oven emis-sions	1910.1044 – 1,2-Dibromo-3-chloropropane (DBCP)
1910.1045 – Acrylonitrile	1910.1047 – Ethylene oxide
1910.1048 – Formaldehyde (formalin)	1910.1050 – Methylenedi-aniline
1910.1051 – 1,3-Butadiene	1910.1052 – Methylene chloride

1926.62 – Lead	1926.1110 – Benzidine
1926. 1112 – Ethyleneimine	1926.1113 – Beta-Propiolac-tone
1926.1144 – 1,2-Dibromo-3-chloropropane	1926.1148 – Formaldehyde

Advantages

- Hazard identification is simple, is the chemical present?
- If you don't have it, the regulation is not applicable.

Disadvantages

- Some overlap between 1910 and 1926 regulations, which may cause confusion.
- 1910 speaks to facility operation, while 1926 speaks to construction therefore the implementation is different. Caution should be used to select the most appropriate standard on mission activities and apply consistently.

Additional Hazard Evaluation (AHE)

Many DOE sites use an additional hazard evaluation (AHE) due to the possibility of the mixing of chemicals or incompatible chemicals that could cause violent exothermic chemical reactions such as a detonation (explosion) or deflagration. An unplanned mixing of chemicals could be the result of mechanical failure or human error such as the introduction of an incorrect feedstock. For example, adding nitric acid to a process designed for sulfuric acid or adding 70% nitric acid where 25% nitric acid was required can result in off-normal conditions. The consequences of mixing could include a rapid temperature rise, toxic gas release, fire, deflagration or detonation. A method for determining whether or not a chemical is incompatible should be developed as a tool to assist in reducing the possibility of inadvertent mixing of incompatible chemicals.

Advantages/Disadvantages

- Identify chemicals that may have incompatibility for proper storage and handling. Process knowledge should be used for chemical mixing and associated hazard assessment for these chemicals. Savannah River Site (SRS) in its WSRC-IM-97-9 manual[4] cites a comprehensive listing of numerous incompatible chemicals.
- If process knowledge is not used in chemical mixing, inadvertent mixing of chemicals may result in:
  a.   Heat generation;
  b.   Fire;
  c.   Deflagration;
  e.   Detonation (Explosion);
  f.   Violent exothermic reaction;
  g.   Toxic fumes.

Non-chemical hazards such as mechanical equipment failure, wrong concentration of a material or leak in a system, etc., can trigger chemical hazards that should also be considered in an AHE..

COMMON HAZARDS SCREENING CRITERIA

Screening criteria

Common characteristic properties of hazardous chemicals are usually NFPA ratings; toxic, corrosive, reactive, ignitable, and incompatible chemicals. Thresholds that may be used for screening include

- RQ 40 CFR 302
- TPQ 40 CFR 355

- TQ 29 CFR 1910.119
- TQ 40 CFR 68

The chemicals that do not screen out can be further evaluated for hazard and accident analysis, either qualitatively or quantitatively, and the selection of controls.

All hazards below the screening criteria should be evaluated by the techniques listed in the ISM. Chemicals not appearing on the RQ list should be checked for the hazard characteristics in the TPQ and TQ, and chemical industry references such as Sax' "Dangerous Properties of Industrial Materials" or the National Institute for Occupational Safety and Health (NIOSH).

Advantage

- Using the proper RQ, TPQ and TQ values for screening, the facility can be classified accordingly and hazards can be further analyzed with graded approach and appropriate controls.

Disadvantage

- Some chemicals do not have a published RQ or TPQ or TQ values for screening, which may increase the difficulty in classifying the facility and hazards, even with graded approach.

Physical Hazards

There are other common facility or process hazards such as pressure, temperature, and voltage, that may be screened out. However, they can serve as initiators for accidents involving chemical hazards. Flammable materials, leaking of materials, and equipment failure are other examples of common hazards, which can serve as initiators for accidents. The following table provides some examples:

Chemical of Concern

Hazard	Screening Criteria
Asphyxiant	Oxygen <19.5%
Explosive	Class A, B, C in 49 CFR 173
Flammable	NFPA Class I or II
Pressure	>3,000 psig
Temperature	Can act as an initiator: Exceeds flash point, volatilize low vapor pressure chemical, increase pressure

FACILITY CHEMICAL HAZARD CLASSIFICATION (CHC)

Cancelled DOE-O-5481.1B and DOE-EM-STD-5502-94 provided guidance on facility chemical hazard classification (CHC) (e.g., high/moderate/low), criteria for categorization (consequence, inventory), safety analysis details, and approval authority. Although many DOE/NNSA sites are still following the same protocols based on their earlier practices or these directives may be still in their contract terms, currently there is no DOE directive or guidance for the facility CHC, screening criteria, selection of controls, level of safety analysis, and approval authority. Each site is following its own protocol of chemical safety analysis practices negotiated with the local field or site office.

Two approaches are viable in the DOE/NNSA complex: 1) industry standard – OSHA (PSM) and EPA (RMP) regulations that do not require traditional facility hazard classification; and 2) traditional CHC that is based on inventory or consequence. Both approaches are discussed as follows.

Industry Standard (OSHA – PSM; EPA – RMP)

DOE/NNSA sites are required to follow the CFR regulations of OSHA and EPA and their use may be required through State Facility Agreement (agreement between State and DOE/NNSA). A site can select an approach suitable to its depth of analysis pertinent to meet the requirements of applicable regulations such as 40 CFR 68 (RMP) and 29 CFR 1910.119, TQ for process safety management (PSM), 40 CFR 355, TPQ for emergency planning and notification, and 40 CFR 302, RQ for spill control for reportable quantities and notification and clean up.

Some DOE sites find that these three layers of control of chemicals addressing environmental, emergency response, and safety provide sufficient controls to identify chemical hazard and that the greater challenge is to consistently apply the controls. These regulations do not require facility CHC, which is an advantage with this approach. However, this approach should be in concurrence with the field or site office of DOE /NNSA.

The OSHA PSM is an industry standard for industrial hazards and focuses mainly towards workers (~100 m). However, it may also be used as part of the DOE's ISMS. The PSM has 14 elements that are geared towards safety management of facilities, operations, technologies, and personnel. These 14 elements are described as follows:

- Employee participation
- Process safety information
- Process hazard analysis (PrHA)
- Operating procedures
- Training
- Subcontractor safety
- Pre-start up safety review
- Mechanical integrity
- Non-routine work authorization

- Management of change
- Incident investigation
- Emergency planning and response
- Compliance audit
- Trade secrets.

The PSM rule is a performance-based regulation; it does not prescribe how each element is to be implemented. Two DOE handbooks (DOE-HDBK-1100-2004 and DOE-HDBK-1101-2004) have been developed to suggest approaches to effectively implement the 14 elements. This section focuses on the process hazard analysis (PrHA, Element #3). If a chemical inventory exceeds the 29 CFR 1910.119 (PSM) TQ, then it is a PSM facility and a PrHA can be performed using techniques such as

- What–If/Checklist or analysis
- Hazard and Operability (HAZOP) analysis
- Failure Mode and Effects Analysis (FMEA)
- Fault Tree Analysis (FTA)
- Event Tree Analysis (ETA)

These techniques are discussed in "HAZARD ANALYSIS." The PrHA typically identifies hazards, assesses hazards of the process, examines causes and consequence of potential accidents, and identifies engineered and administrative controls. The selection of controls is usually based on risk (product of frequency and consequence) rather than on either likelihood of occurrence (frequency) or severity of consequence (DOE-HDBK-1100-2004, Section 3.2.8). The PrHA is *qualitative* (see Table 1, "HAZARD ANALYSIS"). The PSM focuses mainly on worker safety.

Table 1: An Example of a Hazard Evaluation Table (Qualitative HA)

Event No.	Event Category	Hazard	Event Description/Consequence	Causes	Existing Controls Preventive (P) Mitigative (M)
1	Fire	Flammable material; toxic release	Medium fire.	Miscellaneous combustibles, hydrogen from uninterrupted power source battery, and ignition sources. Electrical short. Thermal energy from electrical equipment. Friction from belts.	Design:
			In backpulse Chamber Areas results in release of toxic smoke or gases.		Electrical equipment, P
			Worker injury onsite-1, onsite-2, and offsite exposure.		NFPA standards, P
					Fire detection. and suppression, M
					Building ventilation, M
					Administrative:
					Combustible material control, P
					Trained personnel, P, M
					Standard operating procedure, P
					Fire Dept. response, M
					Emergency Operation Procedure, M
2	Acid spill	Acid release	Nitric acid spills when a holding tank ruptures.	Human error.	Berm, P

			Worker injury and floor damaged, and onsite release.	Equipment failure.	Personal Protective Equipment, M
					Trained personnel, P, M
					Emergency Operating Procedures, M
3	Explosion	Flammable gas	Flammable gas detonation in Lab area, while working with filtrate solution (50 gal) of toxic material, leading to an explosion.	Explosive material: Oxygen diffuses into vapor space and mixes with flammable gas (e.g., benzene) and ignition sources. Electrical short. Thermal energy from electrical equipment. Friction from belts.	Design:
			Onsite burns and worker injury and off-site exposure.		Hood design, P
					Nitrogen supply, P
					Fire detection & suppression, M
					Building ventilation, M
					Administrative:
					Combustible material control, P
					Trained personnel, P, M
					Emergency Operating Procedures, M

The PSM program evaluates and analyzes all process hazards and provides the needed set of controls to protect the worker. The requirements in terms of safety analysis are not extensive for a PSM facility. The format and content of a safety document and approval authority should be negotiated with the DOE/NNSA field or site office.

The 40 CFR 355 TPQ, *Emergency Planning and Notification*, and 40 CFR 302.4 RQ, *Designation, Reportable Quantities,* and Notification, when coupled with an institutional chemical management program, industrial hygiene program, worker safety program, and ES&H program with controls in place are adequate to meet the regulatory requirements to protect the public.

The 40 CFR 68 TQ, Accidental Release Prevention Requirements: Risk Management Programs Under Clean Air Act, requires the submittal of a single RMP that analyzes the worst case release scenario for regulated substances at site boundary (public) that exceed their TQs. The TQs in 40 CFR 68 are usually higher than TQs in PSM. The format and content of a safety document and approval authority should be negotiated with the DOE/NNSA field or site office.

If a site adopts the PSM Rule, the PrHA is primarily qualitiative and then qualitative or quantitative evaluations of frequency, consequence, and risk binning are not required. However, if a site adopts traditional CHC based on inventory or consequence criteria, then qualitative or quantitative evaluations of frequency, consequence, and risk binning may be applicable depending on the type of facility (e.g., high/moderate).

Advantages

- PSM does not require CHC such as High/Moderate/Low.
- Does not require quantitative evaluation of frequency, consequence and risk.
- Controls are usually based on risk, and requirements focuses primarily towards worker.
- Safety document requirement is short.
- May not require DOE field office approval, however, this should be negotiated.

Disadvantages

- There is no hierarchy in hazard classification to better define a facility.

Hybrid Criteria

Both inventory and consequence criteria may be used to determine CHC. For example, the intial CHC can be based on inventory criteria, while the final CHC can be based on consequence criteria (ERPG-3, -2, or -1).

The ERPG/TEEL guidelines are used at Los Alamos National Laboratory (LANL), Lawrence Livermore National Laboratory (LLNL), Oak Ridge-Y12, Pantex, Rocky Flats Environmental Technology Site (RFETS), and West Valley because of the short and variable site boundary distances. An additional consideration in favor of consequence is that for many chemicals, threshold quantities are not listed in EPA or OSHA documents. However, TEELs are listed for more than 2,520 chemicals on the DOE website (Rev. 20, April 2004), which makes it lot easier to use the consequence criteria to determine the CHC. The website is http://www.eh.doe.gov/chem_safety/teel.html.

Note: EPA is currently developing Acute Exposure Guidelines Levels (AEGLs), which are based on five emergency exposure periods (10, 30, and 60 min, 4 hr and 8 hr) and three severity levels (AEGL-1,-2,-3). It is anticipated that ERPGs values may be replaced by AEGL values. The specific AEGL to be used is the 60-minute AEGL; particular levels, such as AEGL-3 and AEGL-2 are the same as ERPG/TEEL-3 and -2. See http://www.orau.gov/emi/scapa/teels.htm.

Advantages

- There is a hierarchy in CHC to better define a facility based on inventory or consequence criteria.
- The level of controls can be better selected based on the CHC to protect the workers and public.
- Quantitative consequence exposure can be evaluated for the worker and public.

Disadvantages

- Safety document requirements for High and Moderate CHC may be more extensive than in PSM requirement by OSHA.
- Some DOE/NNSA sites require approval for all CHC.
- Quantitative Consequence for <100m may not be reliable unless the ARCON96 code is used for dispersion calculations.
- ERPGs/TEELs address only toxicity and may not take into account other chemical and physical hazards (e.g., flammability, deflagration, detonation).

HAZARD ANALYSIS

Hazard analysis (HA) provides a structured approach for evaluation of those process-related, NPH, and man-made hazards from non-nuclear facility activities that potentially could impact facility workers, collocated workers, and the public.

Hazard analysis systematically identifies facility hazards and accident potentials, providing these assessments through hazard identification and hazard evaluation techniques. The HA addresses the credible range of hazards and accidents anticipated for a facility. Typically, a qualitative approach is used in HA to support non-nuclear facilities SB development, including specifically addressing the protection of workers and the public and providing for defense in depth.

There are different approaches to hazard analyses. A graded approach may be useful (see the ISMS guiding principles). It is important that all hazards are analyzed one way or another and the process is systematic and consistent. For hazards that are common in industry (often called standard industrial hazards), consensus standards such as OSHA and EPA standards dictate necessary hazard controls. DOE-unique hazards or common hazards resulting in the release of significant quantities of material or unique applications, or hazards that could initiate an event of significant consequence should be the primary focus of hazards analyses. A screening process may be useful to identify hazards needing detailed analysis.

Chemical hazards addressed in hazard analyses may include toxicological, flammability, explosive, reactive, and other hazardous aspects. Each identified hazard is evaluated to characterize relative risk (i.e., in terms of consequences and expected frequency) of unmitigated hazard scenarios. These analyses can also include a preliminary identification of control options that would prevent or mitigate a malfunction or an upset condition that leads to accident occurrences.

Comparison of Industry and DOE-STD-3009 Approaches

Section 2.5 of the Phase 1 report (CSTC 2003C)[1] shows that several methods are used across the DOE complex to perform hazard analyses. The methods used generally fall into one of two categories: a) a chemical industry approach and b) an approach based on DOE-STD-3009 for nuclear facilities. These approaches are discussed below:

Chemical Industry Approach

The primary references of the chemical industry for hazard evaluation are the PSM approach, and the Center for Chemical Process Safety (CCPS) book *Guidelines for Hazard Evaluation Procedures, Second Edition with Worked Examples* (AIChE, 1992). [6] The PSM standard is often used by the chemical industry as good practice even for facilities that fall below the TQs of highly hazardous chemicals. The PSM standard lists six hazard evaluation techniques, although it allows other equivalent methodologies. The CCPS book describes in detail the six listed PSM methods, plus six additional methods are also described. It points out that some of the methods are "broad brush" techniques most useful early in the design process, others are good for detailed analysis, and still others are applicable to special situations. A number of the techniques focus on developing a list of recommendations for improvements to the process or facility. Several of the techniques suggest identifying "safeguards", which are engineered or administrative controls that prevent or mitigate the hazards.

Advantages

- This method provides consistency with the chemical industry approach, which may be easier to implement for contractors whose workforce has come largely from private industry.
- The analysis may be simpler and require fewer resources than using the DOE-STD-3009-like approach.

Disadvantages

- The analysis may not identify safeguards, and may not identify which are the most important controls. It will likely not analyze the ability of important controls to perform identified safety functions.
- There may be hazards that have not been recognized.

DOE-STD-3009-like Approach

This approach uses the basic methods for hazard evaluation as established in DOE-STD-3009, which starts the same as the chemical industry approach: by picking a hazard evaluation methodology from the chemical industry. Then accidents that can cause release of hazardous materials or energy are analyzed. This analysis includes a qualitative estimation of the frequency and consequences of each event and a listing of engineered systems and administrative controls that would prevent or mitigate the scenario. Typically, the frequency and consequences are both estimated as *unmitigated*, which is before controls are applied. A best practice is to also estimate *mitigated* frequency and consequences, which is after controls are applied, to show the effectiveness of controls for potential accidents that affect both the worker and the public. Engineered systems and administrative controls that significantly contribute to preventing an accident or reducing its consequences may be identified for special treatment to ensure they will perform their safety functions when needed. A further extension of this method used by some sites includes binning hazard scenarios by risk (considering both frequency and consequences) to identify scenarios that require more detailed analysis.

DOE-STD-3009 does not specify which hazard evaluation methodology to use. Instead, it refers the reader to the American Institute of Chemical Engineers, CCPS, *Guidelines for Hazard Evaluation Procedures, Second Edition with Worked Examples* (AIChE, 1992). [6] This reference is cited by DOE-STD-3009 Change Notice 3 as applicable to hazards analysis at non-reactor nuclear facilities and is considered appropriate for use at non-nuclear facilities. An appropriate hazard analysis technique can be chosen from several available standard methods that are widely used by government and industry, as described in the CCPS guidelines.

Advantages

- The analysis identifies safeguards, and supports identifying which are the most important ones. It also supports analyzing the ability of important controls to perform identified safety functions.
- The method has well defined binning of frequency, consequence, and risk rankings to establish with the level of rigor needed.
- The method is consistent with the SB approach for nuclear facilities, so contractors can use the same basic approach to perform hazard analyses for nuclear and non-nuclear facilities.

Disadvantages

- DOE-STD-3009-like analysis may be more structured and complex and require more resources than using the chemical industry approach.
- DOE-STD-3009-like approach uses other nuclear standard and nuclear terminology, which clouds the compliance issues.
- The airborne release fraction/release fraction (ARF/RF) values for chemicals and other hazardous compounds may not be available or applicable in DOE-HDBK-3010 (see "Atmospheric Transport and Dispersion Model").

Hazard Analyses Methodologies

Hazard analysis is used to evaluate identified hazards within the context of the facility and authorized processes. References such as *Guidelines for Hazard Evaluation Procedures* provide the guidelines for selecting hazard evaluation techniques as well as general methodology for completing these techniques. An application of a graded approach in conducting hazards analysis is based on the guidance of DOE-STD-3009, as well as the judgment and experience of the analysts, resulting in the selection of an appropriate hazard analysis technique. The graded approach, as presented in DOE-STD-3009, recommends using methods in proportion to the risk involved to evaluate hazards. A graded approach can use a binning matrix as an adjunct to the hazards evaluation method(s). The aim of the qualitative binning method is to select an appropriate bin in the matrix for a given accident scenario. Use of the bin qualitatively identifies the associated relative risk for a given scenario and then allows for the selection of higher risk scenarios for evaluation of preventive and mitigative controls. Some examples of risk binning matrix are shown in Table 8 of the Phase I report (CSTC 2003C). [1]

The chosen hazard evaluation method should help the analyst to further discriminate the importance of hazards, initiating events, and subsequent controls. Each of these methods will basically result in an initial listing of the hazard and associated consequences. To support the analysis of these hazards, a qualitative assessment of the frequency and likelihood of these consequences should be conducted. Under the chemical industry standard, risk assessment may be used to accomplish prioritization.

Some types of acceptable methods for HA, as provided by the 29 CFR 1910.119 OSHA PSM for process hazard evaluation and the CCPS *Guidelines for Hazard Evaluation Procedures,* [6] include:

- What If/Checklist, combination of What If and Checklist
- HAZOP (Hazard and Operability) Analysis
- FMEA (Failure Modes and Effects Analysis)
- FTA (Fault Tree Analysis)
- ETA (Event Tree Analysis)

Discussions on the application of these methods are provided in the PSM and CCPS references, as well as in the *System Safety Analysis Handbook* (published by the System Safety Society), and training course material (Course # 139, 2002) [7] from ABS Consulting Process Safety Institute. A summary of the above cited methodologies is presented below;

What–If/Checklist

This is one of the most popular methodologies used in hazard analysis. The typical nine steps in What–If/Checklist methodology are shown in Figure 2. Each methodology step is further discussed:

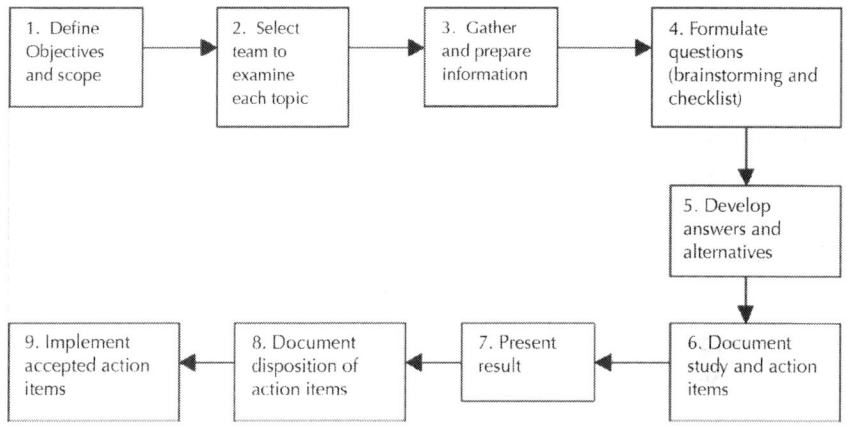

Figure 2: What–If/Checklist Methodology.

What–If Methodology

The purpose of the What–If Methodology is to identify hazards, hazardous situations, or specific accident events that could produce an undesirable consequence. The What–If technique is a loosely structured brainstorming approach in which a group of experienced individuals familiar with a process ask questions or voice concerns about possible undesired events in the process. It is inherently not as structured as some other techniques, such as the HAZOP or FMEA. Rather, it requires the analysts to adapt the basic concept to the specific application.

The "What–If" Analysis concept encourages an analysis team to think of questions that begin with "What If." Through this questioning process, an experienced group of individuals identify possible accident situations, their consequences, and existing safeguards, then suggest alternatives for risk reduction. The potential accidents identified are neither ranked nor given quantitative implications. The analysis team reviews the process from raw material to final product. At each step they ask "what if" questions dealing with procedural errors, hardware failures, and software errors.

The "What–If" Analysis technique may simply generate a list of questions and answers about the process. However, it usually results in a tabular listing of hazardous situations with "What–If", causes, their consequences, safeguards, and possible options for risk reduction for the workers and public.

Checklist Methodology

In a traditional Checklist Analysis, the analyst uses a list of specific items to identify known types of hazards, design deficiencies, and potential accident situations associated with common equipment and operations. The identified items are compared to appropriate standards. The Checklist Analysis technique can be used to evaluate materials, equipment, or procedures. Checklists are most often used to evaluate a specific design with which a company or industry has a significant amount of experience, but they can also be used at earlier stages of development for entirely new systems or processes to identify and eliminate hazards that have been recognized through years of operation of similar systems. This can be done in a tabular form.

Advantages

- Universally applicable to process and non-process issues.
- Can be performed at any design stage.
- Can easily focus on specific concerns (e.g., spill, fire, deflagration, detonation).
- Easy to learn and apply.
- Efficient method.

Disadvantages

- Highly dependent on team experience and/or appropriateness of checklists (s).
- Has potential to miss some meaningful scenarios.

- Difficult to audit for thoroughness.
- Difficult to ensure regulatory compliance (if the what–if technique is used alone).

Hazard and Operability (HAZOP) Analysis

A HAZOP is a systematic examination of all possibilities to identify and assess the significance of the facility SSCs and processes that can malfunction or be improperly operated. Basically, HAZOP analyses are designed to identify potential process hazards resulting from system interactions or exceptional operating conditions.

The study is performed by a multidisciplinary team to identify hazards and operational problems that could result in accident scenarios. The HAZOP team, as identified in 29CFR 1910.119, consists of a team leader with HAZOP experience, a systems engineer with knowledge of facility systems, and a process engineer or operator with intimate knowledge of the process. The size of the HAZOP team will vary according to the scale and complexity of the process.

A HAZOP study relies greatly on design documentation such as piping and instrumentation diagrams (P&IDs), process flow diagrams (PFDs), system design documents, procedures, and equipment and material specifications. In order to perform a successful HAZOP study, it is imperative that the facility and process documentation is up to date and accurate.

The study uses a structured guide word approach to evaluate deviations from normal or design operating parameters such as temperatures, pressures, and flowrates. Guide words such as *none*, *more*, and *less* are applied to the facility and process parameters. For example, applying the guide word *more* to the pressure variable of a facility vessel would result in the operating deviation of increased pressure. The HAZOP team would then determine the possible deviation, causes, consequences, controls, and any suggested actions to reduce or mitigate the risk; the results of which are recorded in a HAZOP table.

Advantages

- Offers a creative approach for identifying hazards, particularly those involving reactive chemicals.
- Thoroughly evaluates potential consequences of process upsets or failure to follow procedures.
- Systematically identifies engineering and administrative controls and consequences of their failures.
- Provides a good understanding of the system to team members.

Disadvantages

- Requires a well-defined system of engineering documentation and procedures.
- HAZOP is time consuming.
- Requires trained engineers or SMEs to conduct the study.
- HAZOP focuses on one-event causes of deviations or failures.

Failure Modes and Effects Analysis (FMEA)

An FMEA is a systematic method for examining the effects of component failures on system performance. Basically FMEA focuses on failures of systems and individual components and examines how those failures can impact facility and processes. FMEA is most effective when a system is well defined and includes the followings key steps:

- Listing of all system components;
- Identification of failure modes (and mechanisms) of these components;
- Description of the effects of each component failure mode;
- Identification of controls (i.e., safeguards, preventive and mitigative) to protect against the causes and/or consequence of each component failure mode;

- If the risks are high or the single failure criterion is not met.

A FMEA table consists of the above five steps: (1) component description; (2) failure mode; (3) effects; (4) controls; and (5) any suggested action. FMEA explores single component/human failure. Multiple FMEAs may be needed to identify hazards in each system configuration (e.g., start up, operations), but Fault Tree Analysis is a better choice for multiple component failures.

At a minimum, the FMEA team should consist of a FMEA team leader with prior experience performing FMEAs and the system engineer responsible for the system that is being evaluated. More complex projects may require several additional personnel such as line managers, safety analysts, technical experts, and scribes. SMEs may be brought in by the team on an-as needed basis during the FMEA study.

Advantages

- Simple
- Efficient
- Cost effective
- Has quantitative applications

Disadvantages

- Limited capability to address operational interface and multiple failures
- Human error examination is limited
- Missing components are not examined
- Common-cause vulnerability may be missed

Fault Tree Analysis (FTA)

A fault tree is a detailed analysis using a deductive logic model (using Boolean algebra logic) in describing the combinations of failures that can produce a specific system failure or an undesirable event. An FTA can model the failure of a single event or multiple failures that lead to a single system failure. An FTA is often used to generate:

- Qualitative description of potential problems
- Quantitative estimates of failure frequencies/likelihoods and relative importance of various failure sequences/contributing events
- Suggested actions to reduce risks
- Quantitative evaluations of recommendation effectiveness

The FTA is a top-down analysis versus the bottom-up approach for the event tree analysis. The method identifies an undesirable event and the contributing elements (faults/conditions) that would initiate it.

The following basic steps are used to conduct a fault tree analysis:

- Define the system of interest.
- Define the top event/system failure of interest.
- Define the physical and analytical boundaries.
- Define the tree-top structure.
- Develop the path of failures for each branch to the logical initiating failure.
- Perform quantitative analysis (if necessary).
- Use the results in decision making.

Once the fault tree has been developed to the desired degree of detail, the various paths can be evaluated to arrive at a probability of occurrence. Cut sets are combinations of components failure causing system failure (i.e., causing the top event of the tree). Minimal cut sets are the smallest combinations causing system failure.

Advantages

- Allows an analyst to quantify risk associated with a failure
- Allows examination of multiple failures
- Provides easily understood graphical models

Disadvantages

- Requires a skilled analyst. It is an art and also a science
- Focuses only on one particular type of problem in a system, and multiple fault trees are required to address the multiple modes of failure
- Graphical model can get complex in multiple failures

Event Tree Analysis (ETA)

An ETA is an inductive analysis that graphically models, with the help of decision trees, the possible outcomes of an initiating event capable of producing a consequence. The procedure for an ETA is shown in Figure 3.

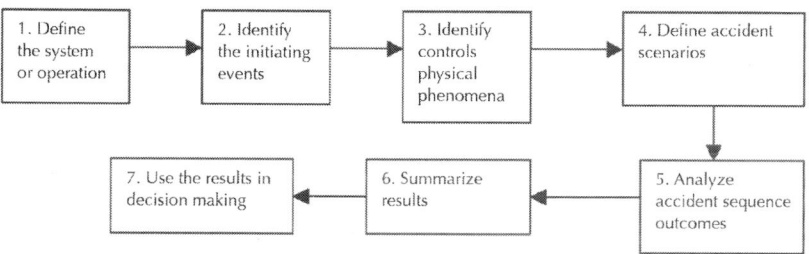

Figure 3: Procedure for Event Tree Analysis.

An analyst can develop the event tree by inductively reasoning chronologically forward from an initiating event through intermediate controls (safeguards) and conditions to the ultimate consequences. An ETA can identify a range of potential outcomes for a specific initiating event and allows an analyst to account for timing, dependence, and domino effects that are cumbersome to model in fault trees.

An ETA is applicable for almost any type of analysis application but most effectively is used to address possible outcomes of initiating events for which multiple controls (lines of assurance) are in place as protective features.

Advantages

- Accounts for timing of events
- Models domino effects that are cumbersome to model in fault trees analysis
- Events can be quantified in terms of consequences (success and failure)
- Initiating event, line of assurance, branch point, and accident sequence can be graphically traced

Disadvantages

- Limited to one initiating event
- Requires special treatment to account for system dependencies
- Quality of the evaluation depends on good documentations
- Requires a skilled and experienced analyst

The above techniques provide appropriate methods for performing analyses of a wide range of hazards during the design phase of the process and during routine operation. A combination of two or three methods (e.g., what–if/checklist and HAZOP) is more useful than individual methods as each method has some advantages and disadvantages. However, some of the more rigorous techniques such as FTA and ETA are reserved for special situations requiring detailed analysis of one or a few specific hazardous situations of concern.

Risk Binning Evaluation

Risk binning is a product of an accident frequency and consequences and is used to rank the risks involved with hazards and activities. Risk binning evaluates these hazard analysis parameters in keeping with the qualitative nature of these analyses. Quantitative measures are typically considered only in special cases.

For nuclear facilities, DOE-STD-3009 states that the purpose of risk binning is "to separate the lower risk accidents that are adequately assessed by hazard evaluation from higher risk accidents that may warrant additional quantitative analysis". A similar approach may be used for non-nuclear facilities. For non-nuclear facilities, risk binning might also be used for grading controls (see "IDENTIFICATION OF CONTROLS") or to determine if appropriate controls are in place to ensure adequate safety.

Analysts may elect to define and analyze unmitigated releases as appropriate for the specification of controls based on ERPG/TEEL-3 or ERPG/TEEL-2 criteria.

Receptors

Immediate workers (10–30 m), collocated workers (typically 100 m), and the public (site boundary) are evaluated for each given scenario. Some scenarios may impact all receptors and some may only impact one receptor. Workers are defined as those within the localized operation or facility area(s) as well as collocated workers within 100 m of the hazard on DOE-controlled premises. The public is defined as people that are outside areas in the direct control of DOE/NNSA. Various examples of immediate workers (onsite-1) and 100 m workers (onsite-2) and the public as adopted by various DOE sites are shown in Table 6 of the Phase 1 report. Some sites combine the immediate worker and the 100-m worker as just the worker.

Consequence

In general, all scenarios have the potential to impact the workers and public. For each scenario, the worst-case consequences are characterized qualitatively to each receptor, using a qualitative consequence matrix. A sliding scale for consequences is applied to the public, while a different scale applies to the localized and onsite worker receptors. A conservative difference typically exists between the consequences for the worker and for the public – e.g., catastrophic for the worker is loss of life, whereas catastrophic for the public is life-threatening injuries. There are different approaches to consequence ranking. Typically, high, moderate, low, and negligible are used based on ERPG/TEEL-3, ERPG/TEEL-2, and ERPG/TEEL-1 criteria (See Table 7 of CSTC 2003-C report).[1]

Frequency

Four frequency (f) levels from an example in DOE-STD-3009 are often used for hazard analysis. These are defined as: Anticipated (AN) – (10^{-1}/y ≥ f ≥ 10^{-2}/y); Unlikely (UN) – (10^{-2}/y ≥ f ≥ 10^{-4}/y); Extremely Unlikely (EU) – (10^{-4}/y ≥ f ≥ 10^{-6}/y); and Beyond Extremely Unlikely (BEU) – (10^{-6}/y ≥ f). The nominal frequency is related to occurrence in the lifetime of the facility. Following the qualitative analysis principles, the nominal frequency should be used as a guide in assigning the relative likelihood for each scenario. A single likelihood ranking is then given for each scenario. The likelihood should be based on subject matter expert (SME) input and need not be based on empirical data. Various examples of frequency rankings are shown in Table 5 of the CSTC 2003-C report. [1]

Examples of a Completed Hazard Evaluation Table

The format of a hazard evaluation table usually reflects the results of the particular hazard evaluation process used, but generally these types of tables provide similar types of information. Hazard evaluation tables typically present a record of identified hazards, causes of events involved, potential consequences, hazard category, and preventive and mitigative control measures. These evaluation tables may be tailored to record a level of results that reflects the rigor provided in the particular hazard evaluation approach.

There may be two general types of hazard evaluation tables: qualitative and semi-quantitative. Examples of completed hazard evaluation tables are shown in Table 1 and Table 2. See Table 1 in "Facility CHC" Section. In a qualitative hazard evaluation table, unmitigated and mitigated frequency/likelihood ranking, consequence ranking, and risk ranking are not included for the worker and public, whereas these parameters are included in a semi- quantitative/hazard evaluation table.

Table 2: An Example of a Completed Hazard Evaluation Table (Semi Quantitative HA)

Event No.	Event Category	Hazard	Event Description/ Consequence	Causes	*Existing Controls* Preventive (P) Mitigative (M)
1	Fire	Flammable material; toxic release	Medium fire. In backpulse Chamber Areas results in release of toxic smoke or gases. Worker injury onsite-1, onsite-2, and offsite exposure.	Miscellaneous combustibles, hydrogen from uninterrupted power source battery, and ignition sources. Electrical short. Thermal energy from electrical equipment. Friction from belts.	Design: • Electrical equipment, P • NFPA standards, P • Fire detection. and suppression, M • Building ventilation, M Administrative: • Combustible material control, P • Trained personnel, P, M • Standard operating procedure, P • Fire Dept. response, M • Emergency Operation Procedure, M
2	Acid spill	Acid release	Nitric acid spills when a holding tank ruptures. Worker injury and floor damaged, and onsite release.	Human error. Equipment failure.	• Berm, P • Personal Protective Equipment, M • Trained personnel, P, M • Emergency Operating Procedures, M
3	Explosion	Flammable gas	Flammable gas detonation in Lab area, while working with filtrate solution (50 gal) of toxic material, leading to an explosion. Onsite burns and worker injury and offsite exposure.	Explosive material: Oxygen diffuses into vapor space and mixes with flammable gas (e.g., benzene) and ignition sources. Electrical short. Thermal energy from electrical equipment. Friction from belts.	Design: • Hood design, P • Nitrogen supply, P • Fire detection & suppression, M • Building ventilation, M Administrative: • Combustible material control, P • Trained personnel, P, M • Emergency Operating Procedures, M

AN: Anticipated; UN: Unlikely; EU: Extremely Unlikely; Risk Ranking: 1 > 2 > 3 > 4; Mod.: Moderate; Neg.: Negligible.

These two types of hazard evaluation tables are essentially modified versions of a Preliminary Hazard Analysis (PrHA) summary worksheet format. The PrHA worksheet format is discussed in Section 6.4 of the AIChE handbook[6] and in MIL-STD-882.

In general, hazard evaluation tables can be tailored to provide those hazard evaluation results that are of interest to and useful for the facility. In addition to hazard evaluation results, some facilities add columns to these tables to denote assignment of follow-on responsibilities and associated schedules to address safety issues, as well as a column for tracking corrective actions implemented by the facility to address safety issues.

CONSEQUENCE/SOURCE TERM ANALYSIS

Introduction

As discussed previously, an HA can be qualitative, semi-quantitative or quantitative. A quantitative analysis may be necessary for higher hazard processes or facilities. A more quantitative analysis is sometimes termed consequence or accident analysis to denote an additional level of rigor than an HA. Accident analyses are also sometimes used to define a design basis event (DBE) for SB purposes. For a quantitative accident analysis, Gaussian dispersion model codes to simulate atmospheric transport and dispersion are commonly used. These models include:

- MACCS2 Model (uses historical meteorological onsite dataset to calculate /Q value)
- Areal Locations of Hazardous Atmosphere (ALOHA)
- Emergency Prediction Information Code (EPIcode)

These models are approved models (codes) by the DOE-EH Central Toolbox Registry (Chung and O'Kula 2002)[8] for safety analysis and are also viable "approved" tool box codes recommended by Safety Analysis Working Group (SAWG)/Energy Facility Contractors Group (EFCOG). DOE-EH has provided computer code application guidance for documented safety analysis for MACCS2, ALOHA, and EPIcode codes in DOE-EH-4.2.1.3, *Code Application Guidance.*

Other chemical consequence models are also used for specific purposes. These are DEGADIS, SLAB, HGSYSTEM, SCREEN3, ARCON96, and ARCHIE. For example, HGSYSTEM and DEGADIS can model heavy gases such as sulfur dioxide and chlorine, where SCREEN3 is not suitable for heavy gases. ARCON96 can calculate concentrations in the vicinity of buildings (short distances) and ARCHIE is used for fire modeling and explosion. The reader should refer to user manuals for these models for additional information. A discussion of 64 consequence assessment models is available from the Office of the Federal Coordinator of Meteorology (OFCM) in "Directory of Atmospheric Diffusion and Consequence Assessment Models". It can be accessed at www.ofcm.gov.

Atmospheric Transport and Dispersion Model

Atmospheric transport and dispersion models that are used for chemical consequence analyses are commonly based on a Gaussian dispersion equation from the *Workbook of Atmospheric Dispersion Estimates, An Introduction to Dispersion Modeling* (Turner 1994): [9]

$$\chi(x, y, z) = \left[\frac{Q}{2\pi u \sigma_y \sigma_z} \right]^{[-y^2/2\sigma_y^2]\{[-(H-z)^2/2\sigma_z^2]-[-(H+z)^2/2\sigma_z^2]\}} \tag{1}$$

where χ is the air concentration, mg/m³; Q, continuous emission rate, mg/s (mass release/time); u, average wind speed, m/s; σ_y, standard deviation of concentration distribution in the crosswind direction (x), m; σ_z, standard deviation of the concentration distribution (function x) in the vertical direction, m; H, the effective release height of the centerline of the plume, m; x, downwind distance, m; y, crosswind distance, m; z: vertical height, m; ϖ, 3.142.

For a ground-level release, $y = 0$, $z = 0$, and $H = 0$. Equation (1) simplifies to

$$\chi(x) = Q(\pi u \sigma_y \sigma_z)^{-1}$$

(2)

$$\frac{\chi}{Q} = (\pi u \sigma_y \sigma_z)^{-1}$$

(3)

χ/Q (s/m³) is the relative atmospheric dispersion for a particular atmospheric condition; and exposure associated with the postulated release to a receptor. Atmospheric stability class (A–F) is a feature to estimate the atmospheric mechanical turbulence and buoyancy for the dispersion in the crosswind (y) and vertical (z) directions downwind (x) from the source. The method may use the Pasquill stability class categories in combination with Pasquill–Gifford dispersion parameters or by dispersion parameters by Briggs (Turner, 1994). [9]

The chemical concentration is calculated by:

$$\text{Concentration } (\text{mg}/\text{m}^3) = \frac{\chi}{Q} \times \text{RR}$$

(4)

where RR is the release rate as mg/s, ST/T and ST, source term; T, release time.

$$\text{ST} = \text{MAR} \times \text{ARF} \times \text{RF} \times \text{DR} \times \text{LPF}$$

(5)

$$\left[\frac{\chi}{Q} \times \text{MAR} \times \text{ARF} \times \text{RF} \times \text{DR} \times \text{LPF} \right] \frac{1}{T}$$

(6)

where $/Q$ (s/m³): Relative atmospheric dispersion for a particular atmospheric condition; typically 50% (median) and 95% meteorology is used. MAR (mg) is the material at risk available for release; ARF, airborne release fraction suspended in air as an aerosol and available for transport; RF, respirable fraction: the fraction of airborne particles that can be transported through air and inhaled into the human respiratory system; commonly assumed to include particles ≤10 μ; Aerodynamic Equivalent Diameter (AED), RF = 1; DR, damage ratio of the total MAR that could be impacted by the accident generated conditions. For a conservative assumption, DR is 1. LPF, Leakpath factor: the fraction of airborne material transported from confinement deposition or filtration mechanism (e.g., fraction of material passing through a HEPA filter); for breach confinement, LPF is 1. T (s): Release duration.

ARF and RF values are usually taken from DOE-HDBK-3010-94 or DOE-STD-1027-92. Release duration is typically 10 or 15 minutes, although a shorter duration (1–3 min) is possible for puff release or small MAR release (e.g., small gas cylinder whose contents are not under pressure). For releases of short duration, a time-weighted average (TWA) of 15 minutes is normally used.

A more detailed treatment of atmospheric transport and dispersion principles can be found in Chapter 9 of the DOE Accident Analysis Guidebook (DOE G 421.1-X) on DOE website – www.directive.doe.gov.

Gaussian distribution (χ/Q Method)

The χ/Q value is a very important meteorological parameter that can vary significantly (1–3 orders of magnitude) depending on meteorological conditions (stability class A to F), thus its accurate determination is crucial. Two approaches to calculate χ/Q values are:

- 95th Percentile: DOE-STD-3009 Appendix A requires the use of Regulatory Guide 1.145 to generate the requisite meteorological data for computing the 95% distribution of concentration or dose to the MOI (maximum offsite individual) or public. This could be considered to be a "worst case" situation as being conservative. The consequence (χ/Q value) is normally obtained through MACCS2 (MELCOR Accident Consequence Code System) by providing a historical meteorological onsite dataset of few years (e.g., 1–5 years of hourly data). If 5 years of data is available, it should be used.

- Persistent Meteorology: For example, a single wind speed and stability class (A to F) is used as input for the duration of the release (e.g., ALOHA, EPIcode, simple hand calculations).

Many sites typically use an F stability class and 1–2 m/s wind speed for initial consequence calculations as being conservative. These codes calculate a centerline Gaussian dispersion plume model as shown in Equation (2). Once a χ/Q value is obtained, then using other parameters listed in Equation (6), chemical concentration (mg/m^3 or ppm) can be hand calculated (including a spreadsheet approach) at a receptor (worker or public) distance.

The χ/Q value is usually not reliable below 100 meters, mainly because of the theoretical model and great uncertainty in the modeling. Therefore, a concentration value for short distance workers (~30 m) is viewed as a qualitative estimate. However, ARCON96 code can be used for short distances.

Aloha and EPIcode

ALOHA and EPIcode are well-developed computer models that can calculate χ/Q values with the weather conditions input provided, such as stability class (A-F), temperature, wind direction, wind height and wind speed, and distance from release. These codes also use a centerline Gaussian dispersion plume model and are user-friendly. ALOHA can model heavy gas releases and has a much more robust evaporation submodel, and can calculate indoor concentrations using infiltration submodels.

With the other information provided as input – e.g., material at risk (MAR), release time, sampling time, receptor height, models calculate concentration (mg/m^3 or ppm) at a given distance (immediate worker, co-located worker, public). These values are then usually compared with ERPG-1, -2, and -3 values, which are based on up to 1-hour exposure. These models are used for gaseous and liquid releases.

A sampling (exposure) time of 15 min. TWA (time weighted average) is recommended to compare with the guideline, which is a conservative estimate for dose assessment to a receptor (Craig et al. 2000).[10] If ERPG-1, -2, and -3 values are not available for a chemical, TEEL-1, -2, and -3 values can be used.[5] Where available, AEGL-1, -2, and -3 values can also be used.

Advantages/Disadvantages

- χ/Q values can be obtained by MACCS2 using historical meteorological onsite dataset of a few years, which is often more reliable method than from a single meteorological conditions at hand by ALOHA and EPIcode or input by hand calculations.

- EPIcode has a feature to print out χ/Q value as a function of distance, where ALOHA does not. EPIcode can select the stability class that maximizes the ground level concentration for elevated release scenarios.

- ALOHA was originally written by NOAA (National Oceanic and Atmospheric Administration) for emergency responder and over the years has been modified to be used in other area. Thus, it has broader applications. ALOHA has a one-hour plume travel limit which truncates analyses at far-field receptors with light wind speeds.

- ALOHA can model heavy gas, whereas EPIcode does not model dense gas releases. ALOHA can model liquid releases from tanks, pipes, and pipelines, whereas EPIcode does not.

- EPIcode was originally written towards emergency preparedness application and now has been broadened towards safety analysis application. Its printout lists all the input parameters and output results, which is in a friendly readable form.

- χ/Q method (MACCS2) and EPIcode have features for deposition velocity in their models, where ALOHA does not.

- In some cases, ALOHA and EPIcode yield reasonably good agreement. In some cases, the models do not; the differences can be attributed to different assumptions or equations in their models (e.g., liquid evaporation model).

- In general, Gaussian based dispersion models yield unreliable results within 100 meters. This may be due to plume meandering or dispersion coefficients that are not suitable for close-in distances. Models have not been validated for use at distances less than 100 m, with the exception of the empirically based ARCON96 code.

- Both ALOHA and EPIcode models are less reliable for conditions of low wind speed or very stable atmospheric conditions.

- Both ALOHA and EPIcode models do not account for building wakes, where MACCS2 accounts for building wake effects.

- MACCS2 is commonly used for dispersion of particulates, although can be used for vapors and gases, whereas ALOHA and EPIcode are commonly used for vapors and gases.

- Both ALOHA and EPIcode can be used as an emergency response tool as a real time in the field, where MACCS2 can not be used as a real time in the field.

IDENTIFICATION OF CONTROLS

The development, identification, and implementation of controls (i.e., engineered and administrative) is an essential step in any safety management process such as ISMS, PSM, or nuclear safety management. Controls are typically based on ERPG/TEEL values and will help to prevent or mitigate analyzed accidents if properly selected, implemented, and maintained. The controls should be based upon the hazard analysis using consequence or risk analysis (usually concurrence with the field or site office). The hazard analysis will identify the scenarios that may require controls. Each accident scenario may have one or more controls to prevent or mitigate the postulated accident. Obviously, accidents with more serious consequences should require more robust controls. The decisions regarding the adequacy of a control set for each accident are made by the hazard analysis team, operating staff, and potentially DOE.

For accidents with minor consequences, the HA team may recommend that safety management programs (SMPs) provide adequate controls. For more serious potential accidents, the team should consider having multiple controls, i.e., defense-in-depth. The team should also prefer engineered controls before considering administrative controls, and preference should be given to preventive over mitigative controls in accordance with applicable DOE guidance and good engineering practice. The defense-in-depth concept also applies to using the safety management programs to increase the robustness of individual engineered controls through regular maintenance and surveillances, configuration management, and training. The identification of controls should include a discussion of the following elements:

- Consideration of any precedence for specific hazard control solutions;
- Identification of engineered controls integral to the design of a facility, equipment, or activity and serving one or more safety functions;
- Identification and description of the devices that measure or monitor a physical condition and notify operators to initiate other actions to shut down the operation, activate another control measure, and/or set off an alarm when a predetermined threshold has been exceeded;
- Identification of the administrative procedures involving personnel who are instructed or trained as appropriate to follow specified procedures;
- Identification of any other activities or measures taken for the purpose of preventing a hazardous situation from developing, or for the purpose of reducing the consequences of that situation, should it occur;
- Identification of safety management programs (SMPs) that provide defense-in-depth to specific administrative or engineered controls;

- Identification of controls to protect initial assumptions and conditions used in hazard and accident analysis.

Grading of Controls

A possible step in the control identification process is the grading of the controls. There is no DOE order or other Federal regulations requiring the grading of controls for non-nuclear, chemically hazardous facilities or activities.

The grading of controls should be performed when justified as increasing safety commensurate with the costs. The process of control grading will rank controls based upon the their significance in reducing the consequence or frequency of a postulated accident. Many different grading schemes can be developed. The benefits of implementing any control grading should be greater than the costs.

Advantages

- The goal of control grading is to provide a more robust and reliable control that will perform its safety function upon demand. Some of the potential benefits of grading include increased emphasis on maintaining and managing the most important controls.
- A simple scheme could select the most important controls that protect workers and the public as safety-related. Another level of control grading could be added for controls that specifically protect the public.

Disadvantages

- Instituting this system within a facility or DOE site may lead to increased costs to develop and maintain the controls and to develop and maintain the grading system. Many DOE sites have active nuclear facilities and the grading scheme could rely upon the nuclear system, minimizing the cost to develop and maintain a separate system.

- Some of the potential costs of grading controls are determining what each control grading level conveys in terms of possible design criteria, determination of control availability, defining safety functions, describing systems, evaluating systems to perform the functions, and selection of potential surveillance or testing requirements. Many of these same issues apply to administrative controls as well.

- There are no written criteria that establish evaluation guidelines, however, there are precedents and established practices within the DOE Complex and industry.

DOE-STD-1186-2004, *Specific Administrative Controls (SACs),* provides additional guidance regarding the use of administrative controls, including SACs that are designated as the principal control for accidents that impact the public or collocated workers. The DOE-STD-1186-2004 is written to address SACs for nuclear hazards; however, the concepts and recommendations can be applied to non-nuclear hazards as well.

Evaluation Guidelines

Before one goes down the path of grading very far, the obvious question arises as to what are the evaluation guidelines for the selection of controls. There are no written criteria that establish evaluation guidelines. However, there are precedents and established practices within the DOE Complex and industry. As shown in Table 9 of the Phase I report (CSTC 2003-C), some sites have developed their own control selection criteria such as ERPG-1, -2, -3 or equivalents for evaluation guidelines. Most sites use EPRG-2 and occasionally ERPG-3 as an evaluation guideline to protect the public and ERPG-3 and occasionally ERPG-2 to protect the collocated worker (100 m). Table 3 lists typical consequence levels and effects.

Table 3: Typical Consequence Levels and Effects

Consequences	Potential Effects
High	Exposures greater than EPRG-3 or 2 (TEEL-3 or 2) or equivalent offsite
Moderate	Exposures greater than EPRG-3 or 2 (TEEL-3 or 2) or equivalent to collocated workers at 100 m.
Low	Significant health effects to local workers (e.g., significant injuries to multiple workers or death)
Minor	Minor health effects to local workers

ERPG-3 is often acceptable for protecting collocated workers due to their hazardous material training, emergency response training, and fitness for duty requirements. EPA's values in the Risk Management Plan (40 CFR 68.130) for protecting the public are based upon ERPG-2 or equivalents (61 FR 31667 et seq, *40 CFR 68, Accidental Release Prevention Requirements: Risk Management Programs Under the Clean Air Act, Section 112(r)(7); List of Regulated Substances and Thresholds for Accidental Release Prevention, Stay of Effectiveness; and Accidental Release Prevention Requirements: Risk Management Programs Under Section 112(r)(7) of the Clean Air Act as Amended, Guidelines; Final Rules and Notice*, June 20, 1996). ERPG-2 or equivalent is also widely accepted as protective of the public within industry and the EPA has implemented it using the rulemaking process.

DOE Complex Practices

As stated earlier, *DOE does not have any requirements to grade controls for chemically hazardous facilities.* However, many DOE sites do grade controls for these facilities. Often, the grading is similar to the practices in use at nuclear facilities and uses terms such as safety significant (SS) or safety features. Additional differentiation could also be added for controls that protect against high, moderate, or low consequences. Each site presently has its own practices.

Chemical Industry Practices

OSHA and EPA do not have any requirements to grade controls for the chemical industry. The concept of control grading is commonly used in the nuclear industry and has not been widely used in the chemical industry. However, as a best practice, some companies in the chemical industry use graded controls (critical vs. non-critical controls) by applying frequency, consequence, and risk criteria. The chemical industry selects the appropriate controls and documents the controls in the required documentation. The regulators expect that the controls will be maintained and controlled appropriately and to take appropriate compensatory actions if the controls are not available. Ultimately, the regulators resort to the General Duty Clause, which obligates an owner to exercise his general duty to protect workers and the public from all types of circumstances, by the installation of additional controls.

Finally, the control identification process can result in the preparation of a safety requirements document. This safety requirements document can be called an operational safety requirements (OSR) document, chemical safety requirements (ChSR) document, or another site-defined term such as work control document (WCD). The safety requirements document will typically define the most important controls for the workers and public that must be maintained to provide a safe operating environment. Controls may be labeled as level 1 for public and level 2 for workers or similar terminology for distinction purpose between the public and workers.

The safety requirements document could list the important active engineered and administrative controls, including surveillance requirements that ensure control availability, other administrative controls including SMP, use and application, and a listing of passive engineered controls. The purpose of the safety requirements document is to provide a concise compilation of controls identified in the hazard analysis for operation of the facility (Table 4).

Table 4: Sample Consequence Levels and Control Preferences

Consequences	Control Preference
High	Engineered control with additional controls providing defense-in-depth. Passive engineered control preferred if feasible. Specific Administrative Controls (SACs) acceptable if DOE-STD-1186[a] is met. SMPs required to protect controls, conditions, and assumptions.
Moderate	Engineered or administrative control with additional controls providing defense-in-depth. Engineered controls preferred if feasible. SACs acceptable if DOE-STD-1186 is met. SMPs required to protect controls, conditions, and assumptions.
Low	Engineered or administrative controls including SMPs. Defense-in-depth approach should be considered if feasible. SMPs required to protect controls, conditions, and assumptions.
Minor	SMPs

[a]DOE-STD-1186-2004 is guidance for developing SACs. It is a requirements document for nuclear facilities only. However, its principles can be applied for non-nuclear facilities.

Preferred operational modes of controls are as follows:

- An engineered control is preferred over an AC.

- Hazard reduction/elimination is preferred over prevention and mitigation.
- A preventor control is preferred over mitigator control.
- A passive control is preferred over active control.
- A preventor control reduces the potential event's frequency (likelihood).
- A mitigitor control reduces the potential event's consequence.

COMMITMENTS TO SAFETY MANAGEMENT PROGRAM

As noted in the DEAR clause of 48 CFR 970.5223-1 and ISMS, the agreed-upon conditions and requirements for safe operation of a facility are requirements of the contract and binding upon the contractor. Development of safety requirements in SB documentation is the process whereby these commitments are established to ensure facility hazards are identified and that controls to prevent and mitigate potential accidents involving those hazards are proposed, approved, and implemented. The safety requirements developed by the contractor and approved by the DOE form a set of commitments to a safety management program (SMP) that are in essence binding for safe operation of a facility.

Commitments can be both engineered safety features and administrative controls. In some cases, more of the safety controls set commitments may be in the form of administrative controls, as opposed to facility engineered design features. As such, an approach may be taken to implement SMP and potentially to use specific administrative controls (SACs) or a similar approach to provide key aspects of the SB for these facilities. In describing these administrative control aspects, it may be important to clearly state those elements and attributes of SMP that are credited in the safety document.

As a minimum, commitments for worker safety and defense in depth identified in the safety document should be covered within relevant SMP (e.g., occupational safety, industrial safety, maintenance, configuration management, quality assurance), as credited in the safety document.

The SMP and related administrative controls could also address other institutional aspects of the safety document, including organization and management, procedures, recordkeeping, assessment, and reporting necessary to ensure safe operation of a facility consistent with the safety requirements committed to by the operating contractor. In general, the administrative controls address:

- Requirements associated with administrative controls, (including those requirements for dispositioning and reporting violations of safety requirement);
- Staffing requirements for facility positions important to safe conduct of the facility;
- Commitments to the SMP identified in the safety document analysis for the facility.

As noted in "IDENTIFICATION OF CONTROLS", controls may be labeled as Level 1 for public and Level 2 for workers or any other terminology for distinction purpose. However, such controls noted as barriers or preventive or mitigative features in the hazard and accident analyses could be addressed in the safety requirements document (e.g., OSR, ChSR).

Requirements for safety function and availability of these engineered features may be addressed though operating limits/ surveillance requirements, SACs, or programmatic safety program commitments. The selection of the particular control approach could be made commensurate with the level of rigor needed to ensure that SB-credited safety functions for these engineered features are met.

The role of programmatic safety commitments could be explicitly stated. The safety document, however, includes only an overview of the program elements and attributes, not the details of the program or its implementing documents. The details of programmatic coverage are not developed in or as part of the safety document.

Discrepancies in the implementation of a program credited in the safety document would not constitute violation unless the discrepancies were so notable as to not provide the elements and attributes of the program that are credited in the safety document.

One overall commitment that could be made in the safety document is that the contractor should not change the facility configuration underlying the documented SB without implementing and completing a review of the change to ensure that new hazards are not introduced, or previously analyzed conditions are not altered. If there is a change or alteration to these set of conditions or parameters, then an unreviewed safety question (USQ)-like process is applicable. The USQ-like process and approval should follow the same protocol as the facility hazard category SB process and approval protocol.

For facilities using PSM/RMP approach, those regulations identify some SMP that may need to be addressed. These are for example, operating procedures, training, management of change, emergency planning and response (see "Industry Standard (OSHA – PSM; EPA – RMP)").

DOCUMENTS AND APPROVAL PROCESS

As noted in DEAR clause of 48 CFR 970.5223-1 and ISMS, the extent of documentation and level of authority for agreement shall be tailored to the complexity and hazards associated with the work and shall be established in a Safety Management System.

The safety document contains the results and discussion of the various steps of the process(es) outlined in various sections such as the SB methodologies, hazards identification, CHC, PrHA, and establishment of appropriate safety controls to protect the workers, public, and the environment.

The level of rigor in the safety documents depends largely on the hazard classification of the chemical facility (e.g., high, moderate, and low; PSM/RMP). The safety document can take various forms using a graded approach such as an auditable safety analysis (ASA), facility use agreement (FUA), hazard control plan (HCP), hazard evaluation report (HER), or other safety document. The document requirement can be negotiated with the local field or site office. Usually, the safety documents are flexible in format but the content should be well defined to address the important steps as outlined above. These practices vary significantly from site to site as noted in Table 24 of the CSTC 2003-C report.

Chemical Hazard Classification (CHC)

Approval of SB documents is provided by the appropriate approval authority. In cases, where DOE sites uses CHC practices (e.g., High, Moderate, Low), typically contractor approval is adequate for a Low hazard facility. For moderate or high hazard facilities, DOE approval of the SB documents may be required, depending on the DOE site specific approval requirements established between the local DOE office and the contractor. The same protocol applies to the USQ-like process for the corresponding High/Moderate/Low type hazard facility.

OSHA and EPA Regulations

If a site selects to follow OSHA 29 CFR 1910.119 and EPA 40 CFR 68 regulations to perform PSM and RMP approaches, the CHC is not required and the PrHA is qualitative (see "FACILITY CHEMICAL HAZARD CLASSIFICATION (CHC)" and "HAZARD ANALYSIS"). The approval and requirements of a document in terms of format, content, depth of analysis, and selection of controls can be short and negotiated with the local field or site office for facilities that are above or below the PSM or RMP.

Approval Process and SER

The review and approval of a SB document are typically negotiated and established on a site-specific basis. Typically, DOE/NNSA approval is required for High and Moderate or PSM/RMP facility. The DOE/NNSA, in its review and approval role, may require modification or addition to the SB commitments made by the contractor. For a formal DOE/NNSA review of a safety assessment, the bases for DOE/NNSA approval are typically documented in a Safety Evaluation Report (SER). This SER may include Conditions of Approval (CoA) that need to be met either prior to implementation of the safety assessment or prior to the next scheduled update of the safety assessment. The SB developers resolve approval issues prior to implementation of the SB or before its next submission, as applicable. The final SER serves as an acceptance of the risk of the operations as described and evaluated in the safety documents by DOE/NNSA.

Advantages/Disadvantages

- PSM/RMP does not require a hazard classification (HC), whereas CHC requires High, Moderate, or Low.
- Safety document requirement can be short for PSM/RMP, where the safety document can be written with a graded approach from High/Moderate/Low.
- PSM process hazard analysis is qualitative, where CHC can be quantitative in some cases as a bounding scenario for High and Moderate HC in the selection of safety controls.
- The approval authority (contractor vs. DOE/NNSA) can be negotiated with the local field or site office for facilities above or below PSM/RMP or depending on the level of CHC - High/Moderate/Low.

RELATED TOPICS

This section discusses two related topics of interest in the development of a chemical, non-nuclear safety document. An EPHA for EMP, which is required by DOE Order 151.1, and explosive and blasting agents required by 29 CFR1910.109 under the purview of 29 CFR 1910.119 are discussed. Some part of the safety document such as HA and controls are applicable to EPHA and explosive areas.

EMERGENCY PLANNING HAZARDS ASSESSMENT (EPHA)

DOE O 151.1 establishes the policy and describes roles and responsibilities for the DOE *Comprehensive Emergency Management System*. The Order requires that the release of or loss of control of hazardous materials be quantitatively analyzed, in an EPHA. If chemicals exceed screening criteria specified in the Order (summarized below), an EPHA must be prepared. Radiological thresholds specified in DOE-O-151.1C are DOE-STD-1027 Category 3 values.

Segmentation of hazardous material inventories is allowed. If the inventory is segregated such that a release could not be caused by a common initiator, each segment may be treated independently. Sealed sources and material packaged in Department of Transportation type B containers is typically excluded from inventory.

If material is in a physical form that makes airborne dispersion unlikely (e.g., particle size >10 μm, or vapor pressure <10 mmHg), it may be excluded. If the material is in the same form, quantity, and concentration as a product packaged for use by the general public, it may be excluded. If the material does not exceed Laboratory Scale (as defined in 29 CFR 1910.1450), it may be excluded. Material with an NFPA 704 Health Hazard Rating <3 may be excluded.

The EPHA uses barrier analysis and normally does not consider frequency. That is, an event should not be dismissed just because it is incredible (beyond extremely unlikely). The analysis may or may not consider barriers (e.g., tank wall) or mitigators (e.g., dike, filter, stack) without regard to functional classification (e.g., SSC, OSR, ChSR).

The spectrum of events requiring consideration in the EPHA is typically greater than that in safety document. For example, minor events that would normally not be considered in a DSA because they are bounded by another event may require analysis in an EPHA because it may lead to a classifiable accident (e.g., Alert, Site Area Emergency). The EPHA must consider malevolent acts as well (although release mechanisms may be the same or similar to events already considered and analyzed). Overall, approaches outlined in this report are applicable here (e.g., "HAZARD IDENTIFICATION", "HAZARD ANALYSIS", "CONSEQUENCE/ SOURCE TERM ANALYSIS", "IDENTIFICATION OF CONTROL", COMMITMENT TO SME, and "DOCUMENTS AND APPROVAL PROCESS"). A realistic worst-case source term is determined. This typically is done for particulates and non-volatile liquids using the DOE Handbook (DOE-HDBK-3010). For evaporative chemical releases, a model such as ALOHA or EPIcode is often used for both source term and consequence assessment for chemical hazards. HOTSPOT is commonly used for radiological hazards. Consequence assessments are typically performed using 95% adverse, or worst case meteorology (see "CONSEQUENCE/ SOURCE TERM ANALYSIS").

Consequences are calculated at specified receptors (30 m, facility boundary, and site boundary) and compared to specified Protective Action Criteria (PAC). For radiological releases, the PAC is 1 rem TEDE (Total Effective Dose Equivalent; 5 rem CEDE thyroid). For chemical release, the PAC is (in order of preference) 60-minute AEGL-2, ERPG-2, TEEL-2).

EPA is currently developing Acute Exposure Guidelines Levels (AEGL-1, -2, -3), which are based on five emergency exposure periods (10, 30 and 60 min., 4 hr and 8 hr) and three severity levels.

It is anticipated that ERPGs values may be replaced by AEGL values. The specific AEGL to be used is the 60-minute AEGL; particular levels, such as AEGL-3 and AEGL-2 are the same as ERPG/TEEL-3 and ERPG/TEEL-2. See http://www.orau.gov/emi/scapa/teels.htm.

The following emergency classes are defined:

Alert: PAC exceeded at 30 m; or 10% of the PAC exceeded at the facility boundary.

Site Area Emergency (SAE): PAC exceeded at the facility boundary

General Emergency (GE): PAC exceeded at the site boundary

Based upon results of the EPHA, Emergency Action Level (EAL) procedures are written to identify conditions that indicate when an emergency classification threshold may have been crossed. In addition, the EPHA documents the technical basis for the Emergency Planning Zone (EPZ). The EPZ must be at least large enough to encompass a circle defined by the distance to the Threshold to Early Lethality (TEL). The TEL is defined by a radiological dose of 100 rem TEDE or a chemical concentration equal to AEGL-3 (or ERPG-3 or TEEL-3).

Advantages/Disadvantages

a. Integrates requirements by other agencies in order to eliminate duplication of efforts.
b. Non-mandatory implementation guidance for this order is published separately in DOE G 151-series Emergency Management Guides.
c. Guidance provides a methodology to examine the potential consequences at distance and develop specific plans and procedures to tailor to the specific hazards present.
d. Protective Action Criteria are well defined for Alert, SAE, GE, and EPZ.

29 CFR 1910, 109 EXPLOSIVES AND BLASTING AGENTS

29 CFR 1910.109, *Explosives and Blasting Agents,* establishes in the Scope section (section (k)(2) of the CFR) that "The manufacture of explosives as defined in paragraph (a)(3) of this section shall also meet the requirements contained in Sec. 1910.119.

Discussion of a Possible Safety Basis Approach

Analysis and control of the hazards associated with the manufacture of explosives must be conducted in accordance with the regulations associated with PSM as defined in 29 CFR1910.119. The methodologies described in this report for development of hazards analysis would also be applicable for the development of hazards analysis for explosives operations.

For operations other than those associated with manufacturing of explosives, 29 CFR 1910.109 does not specifically prescribe SB requirements. However, other drivers (such as ISM and Emergency Planning) may require analysis of these activities as previously discussed in this report.

Clarification of Definition of Manufacture of Explosives

In order to provide clarification to the confusion associated with the definition of the scope of "Manufacture of Explosive," OSHA issued various interpretation letters in response to specific questions from industry. The following summarizes clarifications to the definition of manufacture of explosives are provided based on various OSHA interpretation letters.

Testing, Research Formulation, Evaluation and Analysis

OSHA Interpretation Letter to Mr. F. A. White,[11] Organization Resources Counselors, Inc., states "Activities OSHA considers outside the scope of the explosives manufacturing process if conducted in a separate, non-production research or test area or facility; and do not have the potential to cause or contribute to a release or interfere with mitigating the consequences of a catastrophic release from the explosive manufacturing process include:

- Product testing and analysis which is not part of any in-production sampling and testing of the explosive manufacturing process;
- chemical and physical property analysis of explosive and propellants and pyrotechnic formulations;
- Scale-up research chemical formulations to develop production quantity formulations;
- Analysis of age tests conducted on finished products;
- Failure analysis of tests conducted on pre-manufactured or finished products;
- X-raying;
- Quality assurance testing (not including the extraction of samples from an active explosives manufacturing [production] process);
- Evaluating environmental effects, such as hot, cold, jolt, jumble, drop, vibration, high altitude, salt, and for; and
- Assembly of engineering research and development models.

These operations are covered under the general explosives handling requirements of 29 CFR 1910.109, however, they may require hazards analysis under a separate driver to ensure worker safety. The remainder of the operations involved with the manufacture of explosives is considered to be covered under the scope of PSM.

Nuclear Explosives-Like Assemblies (NELAs) (JTAs/Test Beds/Trainer Assemblies)

OSHA Interpretation Letter to Mr. G. Rountree[12] Aerospace Industries Associated of America, Inc., states "OSHA did not intend that the PSM standard apply to the installation of explosive devices, such as, explosive bolts, detonating cords, explosive actuators, squibs, heating pellets, thermal batteries, ejection seat rocket motors and similar small explosive devices ... into larger finished products or devices that are not intended to explode. The preceding installation is considered a handling activity covered by 1910.109." Based on this interpretation, only those NELAs that are intended to explode or have the potential to cause or contribute to a release or interfere with mitigating the consequences of a catastrophic release (i.e., contain main charge explosives) are covered under the PSM process. This includes all NELAs that contain main charge high explosives. All other NELA-related operations are covered under the general explosives handling requirements of 29 CFR 1910.109, however, they may require an HA under a separate driver to ensure worker safety.

Packaging

OSHA Interpretation Letter to Mr. D. H. Delsemme,[13] August 18, 1994, states "The re-packaging you describe is considered to be storage and handling activities which are not covered by the PSM standard."

Based on this interpretation letter, packaging operations which are not performed as a part of the explosives manufacturing process (i.e., packaging a finished component after completion of a manufacturing related activity) are not covered under the scope of PSM. These operations are under the scope of the general handling requirements of 29 CFR 1910.109, however, they may require HA under a separate driver to ensure worker safety.

Advantages/Disadvantages

    a. Provides a list of specific chemicals and some general categories.
    b. Establishes very specific criteria for manufacturer, storage, transportation and use of explosives and blasting agents.
    c. Speaks only to those chemicals classifiable as explosives or blasting agents.
    d. Would require implementation in conjunction with PSM or RMP for mixed-use facilities.

Other Drivers Associated with Explosives

In addition to the requirements discussed in "29 CFR 1910 109 Explosives and Blasting Agents" above, there are various drivers exist that relate to development of hazards analysis for explosives operations. The Contractors Requirements Document (Attachment 2, items 9 and 10) from DOE O 440.1A, *Worker Protection Management*, requires the contractor to implement a hazard prevention/abatement program to identify, analyze and control hazards in the work place. These hazards would include those associated with explosives operations, and assumes the application of a graded approach for their evaluation and control. As incorporated by DOE O 440.1A, the DOE M 440.1-1, DOE *Explosives Safely Manual* requires an explosives hazards analysis for those facilities where explosives are used, stored or manufactured. This manual specifically references the use of the OSHA defined Process hazard analysis (PrHA) found in 29 CFR 1910.119, Process Safety Management, for any activity involving the manufacturing, formulation, synthesis, testing or disposal of explosives covered by this manual. However, the specific operations for which this requirement applies are not clearly identified, and as such, requires some sight level evaluation, interpretation, and decision to determine which operations are covered.

CONCLUSIONS

This report presents the methods, together with the advantages and disadvantages, for developing a safety document for chemical, non-nuclear facilities. The outline of a non-nuclear hazards analysis document is provided in various steps.

- Facility and Work Description
- Hazard Identification
- Facility Hazard Classification; Industry- PSM/RMP vs. traditional-high/moderate/low
- Hazard Analysis; Qualitative and/or semi-quantitative
- Identification of Controls
- Commitments to Safety Management Program (SMP)
- Document and Approval Process

The outline follows the essential steps of the ISMS as well as incorporates those ideas from DOE nuclear facilities safety document and industry based analyses.

The facilities should discuss the concepts, methods, and strategies with the respective DOE field or site offices to develop the necessary process(es) that ensure protection of the worker, public, and environment from hazardous material releases from high/moderate hazard facilities.

A standard industry approach following the OSHA and EPA (PSM, RMP) requirements and/or an approach similar to the DOE/NNSA nuclear facility SB process (DOE-STD-3009 like) are viable options.

This report isnota proposed standard nor is it guidance for the SB process. This report outlines various safety analysis steps and methodologies with the advantages and disadvantages associated with them, so that each DOE/NNSA site can decide on its own the merits and demerits of each approach. Adoption of any step of the safety document process is voluntarily.

ACKNOWLEDGEMENTS

Twenty-three reviewers provided comments on the report, which were very helpful and greatly improved the quality of this document, and their valuable comments are highly appreciated by the team members. The reviewer's names and their organizations are shown above. The team members also thank technical editing by Barbara Stirrup of Outrider Environmental.

The authors would like to acknowledge the U.S. Department of Energy and their respective organizations for support of this work.

#	Reviewer Name	Organization
1	Jofu Mishima	Consultant, Los Alamos National Laboratory
2	Laurence G. Lee	Idaho National Engineering Laboratory (INEEL)
3	William Von Holle[a]	Defense Nuclear Facility Safety Board (DNFSB)
4	Terry Foppe	Foppe & Associates
5	Deborah Christensen	National Nuclear Security Administration (NNSA), Sandia Site Office
6	Harvey Canter	Lawrence Livermore National Laboratory (LLNL)
7	Jerry C. Bueck	Los Alamos National Laboratory (LANL)
8	L.E. McCurry	ENERGS, Inc
9	Michael E. Cournoyer	Los Alamos National Laboratory (LANL)
10	Marco S. Colalancia	NWIS-NA
11	Jim E. Goss	NNSA-Y-12 Site Office
12	Robert D. Vrooman	NNSA, Sandia Site Office
13	James Fairobent	NA-41, DOE-Headquarters
14	David E. Freshwater	Science Applications International Corporation (SAIC)
15	Tony Villeges	Los Alamos National Laboratory
16	Charlie Satterwhite	Bechtel Jacob, Oak Ridge
17	Mike Harrison	Washington Safety Management Solutions (WSMS)
18	James L. Woodring	Argonne National Laboratory (ANL)
19	Patrice McEahern	CALIBRE
20	David J. Seidel	Los Alamos National Laboratory
21	Adam B. Cohen	Argonne National Laboratory
22	Vishwa Kapila	EH-23, DOE-Headquarters
23	Carl A. Mazzola	Shaw Environmental and Infrastructure, Inc.

a Own view.

Appendix A: Description of Relevant DOE Orders and CFR Regulations

#	Reference	Title	Description
1	DOE-O- 420.1A	Facility Safety	DOE-O-420.1A, *Facility Safety*, establishes facility safety requirements for DOE and NNSA for nuclear safety design, criticality safety, fire protection, natural phenomena hazards mitigation, and a system engineer program. The Order is split applicability for nonnuclear and nuclear facilities as well as explosive facilities. The Order requires that a fire hazards analysis (FHA) be developed for all DOE facilities, nonnuclear and nuclear facilities. The FHA is a comprehensive evaluation of fire hazards in a facility and includes the postulation of fire accident scenarios and estimates of their potential consequences (i.e., maximum credible fire loss). For nonnuclear and nuclear facilities, the Order also requires a Natural Phenomena Hazard (NPH) Assessment.
2	DOE-G 420.1-2	Guide for the Mitigation of Natural Phenomena Hazards for DOE Nuclear Facilities and Non-nuclear Facilities	DOE-G-420.1-2 provides guidance for implementing the natural phenomena hazards (NPH) mitigation requirements of DOE O 420.1, Facility Safety. The guide addresses radiological and nonradiological hazards and life-safety issues, including protection of workers from exposure to hazardous materials that is caused by the failure of structures, systems, and components (SSCs). A contractor or operator responsible for a DOE nuclear or nonnuclear facility must design, construct, and operate the facility so that the public, workers, and environment are protected from the adverse impacts of the listed NPHs.
			The four DOE Standards (DOE STD 1020, 1021, 1022, and 1023) have been developed to provide specific acceptance criteria for various aspects of NPH to meet the requirements of DOE O 420.1, *Facility Safety* and DOE G 420.1-2.
			• DOE-STD-1020, "Natural Phenomena Hazards Design and Evaluation Criteria for Department of Energy Facilities"
			• DOE-STD-1021, "Natural Phenomena Hazards Performance Categorization Guidelines for Structures, Systems, and Components"

8	10 CFR 830, Subpart B	SB Requirements	Subpart B establishes SB requirements for hazard category 1, 2, and 3 DOE nuclear facilities and is not applicable to non-nuclear facilities. The contractor must obtain approval from DOE for the methodology used to prepare the documented safety analysis for the facility unless the contractor uses a methodology set forth in Table 2 of Appendix A to this Part. The documented safety analysis for a hazard category 1, 2, or 3 DOE nuclear facility must, as appropriate for the complexities and hazards associated with the facility as follows:
			• Describe the facility (including the design of safety structures, systems and components) and the work to be performed.
			• Provide a systematic identification of both natural and man-made hazards associated with the facility.
			• Evaluate normal, abnormal, and accident conditions, including consideration of natural and man-made external events, identification of energy sources or processes that might contribute to the generation or uncontrolled release of radioactive and other hazardous materials.
			• Derive the hazard controls necessary to ensure adequate protection of workers, the public, and the environment, demonstrate the adequacy of these controls to eliminate, limit, or mitigate identified hazards, and define the process for maintaining the hazard controls.
			See DOE-G-421.1-2, "Implementation Guide for Use in Developing Documented Safety Analyses to Meet Subpart B of 10 CFR 830" for additional guidance.
9	10 CFR 850	Chronic Beryllium Disease Prevention Program	10 CFR 850, Chronic Beryllium Disease Prevention Program, establishes a chronic beryllium disease prevention program (CBDPP). A baseline inventory is required to identify those areas that contain beryllium and the responsible employer must evaluate potential exposures by performing a beryllium hazard assessment. These assessments should include analyses of existing conditions, exposure data, medical surveillance trends, and the exposure potential of planned activities.

10	29 CFR 1910.109	Explosives and Blasting Agents	For facilities that manufacture explosives, 29 CFR 1910.109, *Explosives and Blasting Agents*, invokes the requirements of PSM (29CFR1910.119), including the completion of a hazards analysis. Two issues should be noted. First, the requirement to use PSM applies to all facilities (including laboratories) that manufacture any amount of explosive. There is no de minimus quantity. Second, the requirement to use PSM does not apply to facilities that store explosives even though 29CFR1910.109 has numerous regulations concerning the storage of explosives in bunkers or other specialized facilities and the structure of these facilities. The PSM standard does not address explosives.
11	29 CFR 1910.119 and 1926.64	Process Safety Management	29CFR1910.119 and 1926.64, Process Safety Management (PSM), has many requirements for the management of industrial chemicals that are listed in these standards. One of these requirements is a chemical process hazard analysis (PrHA) for facilities having listed chemicals present in quantities that exceed threshold quantities for approximately 140 chemicals.
			The PrHA by PSM shares some similarity to the documented safety analysis (DSA) that is required by 10 CFR 830, Subpart B for DOE nuclear facilities. The PrHA and the DSA serve as the primary analysis of facility level hazards, and both involve the following processes:
			• Identification of hazardous material or radionuclide inventories;
			• Implementation of formal hazard analysis techniques that are commensurate with facility complexity;
			• Identification of systems and equipment vital to safety;
			• Formal documentation of findings; and
			• Periodic updates of hazard analysis information.
			Both OSHA PSM and DOE DSA references require the use of established, standard hazard evaluation methodologies. The OSHA PSM requires qualitative PrHA that include What–If/Checklist, Hazard and Operability Study (HAZOP), Failure Mode and Effects Analysis (FMEA), Fault Tree Analysis (FTA), Event Tree Analysis (ETA), and other acceptable methods.

12	29 CFR 1910.120	Hazardous Waste Operations and Emergency Response	OSHA (29 CFR 1910.120) requires that a health and safety plan (HASP) be prepared for hazardous waste cleanup operations. The HASP must involve a hazard/risk assessment of planned activities to identify any conditions that pose significant hazards to workers. A thorough hazard characterization provides the primary basis for the hazard/risk assessment and typically includes a facility walk down, visual inspections, air monitoring and sampling, and a review of facility records. This regulation applies only to clean up at hazardous waste sites; operations at treatment, storage and disposal facilities; or where emergency response operations are anticipated.	
13	29 CFR 1910 and 1926	Various Hazard or Activity Specific OSHA regulations	A number of regulations have hazard analysis requirements that are specific to certain activities, hazardous conditions, or specific substances. These rules include substance or operation specific hazards such as lead, asbestos, beryllium, confined spaces, laboratory operations, and blasting operations. The hazard analysis requirements of this type are an integral part of work planning that feeds into the preparation of hazardous and radiation work permits, Health and Safety Plans, Industrial Hygiene Plans and overall work packages and documentation. These activities have a different emphasis than facility-level hazard analysis, because these are primarily focused on worker protection. As such, activity-level hazard analysis addresses the hazards associated with individual job functions and tasks.	
			For the below listing, these regulations do not specifically provide for hazard analyses or screening quantities, but do detail many requirements for those areas where these chemicals are stored or used. Requirements for regulated work areas, signage, training, etc., should be reflected in the appropriate SB documentation. Many chemicals overlap between 1910 and 1926. Only one regulation is cited for that chemical. These are shown below.	
			1910.1001 – Asbestos	1910.1002 – Coal tar pitch volatiles
			1910.1003 – 13 carcinogens (4-nitrobiphenyl, etc.)	1910.1004 – Alpha-naphthylamine

			1910.1006 – Methyl chloromethyl ether	1910.1007 – 3,3'-Dichlorobenzidine (and its salts)
			1910.1008 – bis-Chloromethyl ether	1910.1009 – Beta-naphthylamine
			1910.1010 – Benzidine	1910.1011 – 4-Aminodiphenyl
			1910.1012 – Ethyleneimine	1910.1013 – Beta-propiolactone
			1910.1014 – 2-Acetylaminofluorene	1910.1015 – 4-Dimethylaminoazobenzene
			1910.1016 – N-Nitrosodimethylamine	1910.1017 – Vinyl chloride
			1910.1018 – Inorganic arsenic	1910.1025 – Lead
			1910.1027 – Cadmium	1910.1028 – Benzene
			1910.1029 – Coke oven emissions	1910.1044 – 1,2-Dibromo-3-chloropropane (DBCP)
			1910.1045 – Acrylonitrile	1910.1047 – Ethylene oxide
			1910.1048 – Formaldehyde (formalin)	1910.1050 – Methylenedianiline
			1910.1051 – 1,3-Butadiene	1910.1052 – Methylene chloride
			1926.62 – Lead	1926.1110 – Benzidine
			1926. 1112 – Ethleneimine	1926.1113 – Beta-Propiolactone
			1926.1144 – 1,2-Dibromo-3-chloropropane	1926.1148 – Formaldehyde

14	40 CFR 68	Chemical Accident Prevention Provisions	The Chemical Accident Prevention regulation requires facilities to meet the planning and analysis requirements of the applicable level of a three level program that increases in stringency. For all three levels, facilities exceeding established thresholds for a limited set of chemicals are required to submit a risk management plan (RMP). The RMP requires analysis of the worst-case release scenario for the facility process(es) to ensure that the nearest public receptor is beyond the distance to a toxic, explosion, radiant heat, or flammable endpoint.
			In addition, a five-year accident history for the processes must be evaluated. For the next two levels of stringency, the RMP must also include documentation that the facilities have implemented a RMP, conducted a hazard assessment, implemented an emergency response program, and developed an accident prevention program. The hazard assessment requires a review of the hazards associated with the regulated substances, process, and procedures. The hazards review identifies the hazards associated with the process and regulated substances; opportunities for equipment malfunctions or human errors that could cause an accidental release; the safeguards used or needed to control the hazards or prevent equipment malfunction or human error; and any steps used or needed to detect or monitor releases.
15	40 CFR 302.4	Designation, Reportable Quantities, and Notification	This regulation designates under section 102(a) of the Comprehensive Environmental Response, Compensation, and Liability Act of 1980 ("the Act") those substances in the statutes referred to in section 101(14) of the Act, identifies reportable quantities for these substances, and sets forth the notification requirements for releases of these substances. This regulation also sets forth reportable quantities for hazardous substances designated under section 311(b)(2)(A) of the Clean Water Act. 40 CFR 302.4, Designation, Reportable Quantities, and Notification, provides a list of hazardous substances and their reportable quantities (RQs). These reportable quantities are those that if exceeded in a release require the notification to the National Response Center and possibly the state in which the release occurred.

16	40 CFR 355	Emergency Planning and Notification	This regulation establishes the list of extremely hazardous substances (EHS), threshold planning quantities (TPQs), and facility notification responsibilities necessary for the development and implementation of State and local emergency response plans. The requirements of this section apply to any facility at which there is present an amount of any extremely hazardous substance equal to or in excess of its threshold planning quantity.
17	48 CFR 970.5204-2 (c)(2)	Laws, Regulations, and DOE Directives	Environmental, safety, and health (ES&H) requirements appropriate for work conducted under DOE contracts may be determined by a DOE approved process to evaluate the work and the associated hazards and identify an appropriately tailored set of standards, practices, and controls, such as a tailoring process included in a DOE approved Safety Management System implemented under the clause entitled "Integration of Environment, Safety, and Health into Work Planning and Execution."
18	48 CFR 970.5223-1	Integration of Environment, Safety, and Health into Work Planning and Execution	For Department of Energy facilities, the primary hazard analysis requirement is found in the DOE Acquisition Regulations (DEAR, ES&H Clause), which requires the identification and evaluation of hazards associated with work as part of an overall documented safety management system (i.e., ISM). The purpose of the ISM is to identify and analyze potential dangers to workers, public or environment to ensure that effective controls can be established to minimize or prevent adverse impacts.
			For additional guidance see DOE G 450.4-1B, "Integrated Safety Management System Guide", March 1, 2001.
19	DOE-G 440.1-2	Locally Enforced Fire/Building Codes	This guide requires that DOE facilities follow numerous codes and regulations including the locally enforced building and fire codes. Every building/fire code used in the United States contains provisions for hazardous materials (e.g., Article 80 of the Uniform Fire Code, Chapter 22 of the Southern Building Code, Chapter 27 of the International Fire Code) These codes require every hazardous material present in the facility to be evaluated to determine all hazards associated with them. Hazards classifications are present in each fire/building code and are similar from code to code.

			Examples of hazards are toxic; highly toxic; class 1, 2, 3 or 4 oxidizer; class I, II, III, IV or V organic peroxide; class 1, 2, 3, or 4 unstable reactive, pyrophoric, etc. If any chemical hazard is present over specified limits in a given facility, then special storage conditions, facility design, and controls to mitigate the hazards may need to be implemented.

DEFINITIONS OF REGULATORY LIMITS AND GUIDELINES

Acute Exposure Guideline Level (AEGL): AEGLs for hazardous substances are being developed by the National Advisory Committee on AEGLs. The AEGLs are based on five emergency exposure periods (10 and 30 min., 1 hr, 4 hr, and 8 hr) and three severity levels as defined below.

AEGL-1: Airborne concentration of a substance above which is predicted that the general population, including susceptible individuals, could experience notable discomfort, irritation, or certain asymptomatic nonsensory effects. However, effects are not disabling and are transient and reversible upon cessation of exposure.

AEGL-2: Airborne concentration of a substance above which is predicted that the general population, including susceptible individuals, could experience irreversible or other serious, long-lasting adverse health effects or an impaired ability to escape.

AEGL-3: Airborne concentration of a substance above which is predicted that the general population, including susceptible individuals, could experience life-threating health effects or deaths.

Emergency Response Planning Guidelines (ERPG) provides values intended as estimates of concentration ranges where one might reasonably anticipate observing adverse effects as a consequence of exposure to a specific substance. Three ERPG values are given in each guide:

ERPG-1: The maximum airborne concentration below which it is believed that nearly all individuals could be exposed for up to 1 hour without experiencing other than mild transient adverse health effects or perceiving a clearly defined, objectionable odor.

ERPG-2: The maximum airborne concentration below which it is believed that nearly all individuals could be exposed for up to 1 hour without experiencing or developing irreversible or other serious health effects or symptoms which could impair an individual's ability to take protective actions.

ERPG-3: The maximum airborne concentration below which it is believed that nearly all individuals could be exposed for up to 1 hour without experiencing or developing life-threatening health effects.

Immediately Dangerous to Life and Health (IDLH): The atmosphere of a work environment that poses an immediate hazard to life or poses an immediate irreversible debilitating effect on health. This term is defined within Occupational Safety and Health Administration (OSHA) regulation Title 29 of the Code of Federal Regulations (CFR) 1910.120, *Hazardous Waste Operations and Emergency Response*.

Permissible Exposure Limit (PEL): Are established by OSHA to protect workers against the health effects of exposure to hazardous substances. PELs are regulatory limits on the amount or concentration of a substance in the air. Some substances may also contain a skin designation. PELs are enforceable and are based on an 8-hour time weighted average exposure.

Temporary Emergency Exposure Limits 1, 2, and 3 (TEEL-1, 2, and 3): Where ERPG – 1, 2, and 3 values are not available, TEEL values can be used. TEEL limits are listed for over 2,520 chemicals. These are alternate guideline limits based on comparisons between toxicity parameters and ERPGs.

Threshold Limit Value (TLV): Guidelines prepared by the ACGIH designed for use in making determinations on the safe levels of exposure to various chemical substances and physical agents found in the workplace. These exposure limits are considered guidelines and are prepared by the ACGIH as best practices in preventing disease or injury.

Integrated Safety Management Systems (ISMS): A Safety Management System to systematically integrate safety into management and work practices at all levels of activity as required by Department of Energy P 450.4, Safety Management System Policy. An ISMS consists of five core functions, which are defined as: 1. Define work, 2. Identify and analyze hazards, 3. Develop and implement controls, 4. Perform work safely, and 5. Ensure performance and continuous improvement.

Occupational Safety and Health Administration (OSHA): Provides regulatory control on exposure limits to chemicals within the work environment quantified as a Permissible Exposure Limit. Regulates the type and quantity of certain listed chemicals to prevent or minimize the consequences of catastrophic releases of toxic, reactive, flammable, or explosive chemicals. These releases may result in toxic, fire, or explosion hazards and are documented in Title 29 of the Code of Federal Regulations Part 1910, subpart 119, Process Safety Management of highly hazardous chemicals and also addressed in Title 29 of the Code of Federal Regulations 1910.120.U.S. Environmental Protection Agency (EPA): Provides for the protection of human health and safeguarding the natural environment. Regulations applicable to the release of hazardous chemicals is covered in Title 40 of the Code of Federal Regulations subpart 68, Chemical accident prevention provisions; 40 CFR 302, Designation, reportable quantities, and notification; and 40 CFR 355, Emergency planning and notification

REFERENCES

1.   CSTC-2003C. J. C. Laul, Phase I Report: Current Hazard Characterization Practices in the DOE Complex. LA-UR-03-1242 (Los Alamos National Laboratory), October 2003.

2.   Laul, J. C. "Some chemical safety aspects at LANL".

Proceedings of EFCOG; 11th Annual Safety Analysis Working Group (SAWG) Workshop. Milwauki, WI, June 21–26, 2001,

3. Cournoyer, Michael E.; Maestas, Marvin M. "Addressing safety requirements through management walkarounds". Chem. Health Safety, 2004, 11(6), 12.

4. Savannah River Site (SRS), WSRC-IM- 97-9 Manual, Rev. 1. March 1999.

5. Craig, D. K.; Lux, C. R. Methodology for Deriving Temporary Emergency Exposure Limits (TEELs), WSRC-TR-98- 00080. Aiken, SC: Westinghouse Savannah River Company, 1998.

6. 6. Center for Chemical Process Safety (CCPS). Guidelines for Hazard Evaluation Procedures, Second Edition with Worked Examples; AIChE; New York, 1992.

7. System Safety Society. System Safety Analysis Handbook and Training Course #139 Material. ABS Consulting, 2002.

8. Chung, D. Y.; O'Kula, K. R. Accident Analysis Guidance for Completion of 10 CFR 830 Compliance DSAs. Oak Ridge, TN: EFCOG, June 22–27, 2002.

9. Turner, Bruce. Workbook of Atmospheric Dispersion Estimates, An Introduction to Dispersion Modeling, 2nd ed. Lewis Publisher, 1994.

10. Craig, D. K.; Davis, J. S.; Lee, L. G.; Prowse, J.; Hoffman, P. W. Toxic Chemical Hazard Classification and Safety Evaluation Guidelines for Use in DOE Facilities, WSRC-MS-92-206, Rev. 3. December 2000.

11. Letter from J. B. Miles, Jr., Directorate of Compliance Programs, OSHA, Department of Labor to Frank A. White, Organization Resources Counselors, Inc. Subject: "Applicability of PSM standard to explosive and pyrotechnic manufacturing." February 4, 1998.

12. Letter from J. B. Miles, Jr., Directorate of Compliance Programs, OSHA, Department of Labor to Mr. G. Rountree, Aerospace Industries Association of America, Inc. Subject: "Process Safety Management Standard." December 2, 1994.

13.   Letter from J. B. Miles, Jr., Directorate of Compliance Programs, OSHA, Department of Labor to Mr. D. H. Delsemme. Subject: "Explosives and Blasting Agents." August 18, 1994.

Citations

CHAPTER 1

Tasneem Abbasi, H.J. Pasman, S.A. Abbasi, A scheme for the classification of explosions in the chemical process industry, Journal of Hazardous Materials, Volume 174, Issues 1–3, 15 February 2010, Pages 270-280, ISSN 0304-3894, http://dx.doi.org/10.1016/j.jhazmat.2009.09.047.

CHAPTER 2

sRobert Zalosh, Dust collector explosions: a quantitative hazard evaluation method, Journal of Loss Prevention in the Process

Industries, Available online 11 March 2015, ISSN 0950-4230, http://dx.doi.org/10.1016/j.jlp.2015.03.011.

CHAPTER 3

Joseph H. Saleh, Amy M. Cummings, Safety in the mining industry and the unfinished legacy of mining accidents: Safety levers and defense-in-depth for addressing mining hazards, Safety Science, Volume 49, Issue 6, July 2011, Pages 764-777, ISSN 0925-7535, http://dx.doi.org/10.1016/j.ssci.2011.02.017.

CHAPTER 4

Q. Kwok, B. Acheson, R. Turcotte, A. Janès, G. Marlair, Fire and explosion hazards related to the industrial use of potassium and sodium methoxides, Journal of Hazardous Materials, Volumes 250–251, 15 April 2013, Pages 484-490, ISSN 0304-3894, http://dx.doi.org/10.1016/j.jhazmat.2013.01.075.

CHAPTER 5

Attri P, Baik KY, Venkatesu P, Kim IT, Choi EH (2014) Influence of Hydroxyl Group Position and Temperature on Thermophysical Properties of Tetraalkylammonium Hydroxide Ionic Liquids with Alcohols. PLoS ONE 9(1): e86530. doi:10.1371/journal.pone.0086530.

CHAPTER 6

Florian Ettner, Klaus G. Vollmer, and Thomas Sattelmayer, "Numerical Simulation of the Deflagration-to-Detonation Transition in Inhomogeneous Mixtures," Journal of Combustion, vol. 2014, Article ID 686347, 15 pages, 2014. doi:10.1155/2014/686347.

CHAPTER 7

Mohammad Hossein Ordoueia, Ali Elkamela, Ghanima Al-Sharrahb, New simple indices for risk assessment and hazards reduction at the conceptual design stage of a chemical process, doi:10.1016/j.ces.2014.07.063

CHAPTER 8

J.C. Laul, Fred Simmons, James E. Goss, Lydia M. Boada-Clista, Robert D. Vrooman, Rodger L. Dickey, Shawn W. Spivey, Tim Stirrup, Wayne Davis, Perspectives on chemical hazard characterization and analysis process at DOE, Journal of Chemical Health and Safety, Volume 13, Issue 4, July–August 2006, Pages 6-39, ISSN 1871-5532, http://dx.doi.org/10.1016/j.chs.2005.06.001.

Index